Study Guide to accompany

Chemistry
& Chemical Reactivity

THIRD EDITION

Kotz & Treichel

Harry E. Pence

State University of New York
at Oneonta

Saunders Golden Sunburst Series
Saunders College Publishing
Harcourt Brace College Publishers

Fort Worth Philadelphia San Diego New York Orlando Austin
San Antonio Toronto Montreal London Sydney Tokyo

Banks; Student Solutions Manul to accompany Chemistry and Chemical Reactivity, 3E. Kotz and Treichel.

ISBN 0-03-001308-9

67 021 765432

TABLE OF CONTENTS

ACKNOWLEDGEMENTS

Many individuals have helped me during the development of the three editions of this study guide, and I'm happy to take this opportunity to express my appreciation to them. Jack Kotz generates enthusiasm in everyone whom he encounters, and it was a pleasure to work on this project with him. I wish to thank John Vondeling and Kate Pachuta for their patience and assistance during the writing of the first edition and Sandi Kiselica for her work on the second edition. Beth Rosato provided similar support for the third edition. Susan Harman deserves special mention for the invaluable contribution she has made to the second and third editions.

It's impossible to acknowledge all of my colleagues and former teachers who have contributed to this endeavor, but I do wish to specifically thank Larry Armstrong and Bruce Knauer from the Chemistry Department at SUNY Oneonta. A special note of gratitude is necessary for the many students who have helped me to better understand the difficulties of learning chemistry and have also suggested ways to improve this study guide.

The armchair exercises have been developed from a wide variety of sources, and whenever possible the origin is indicated. Probably in some of the other cases the exercise originated from some now-forgotten resource, either a personal contact or something read in the literature. Therefore it is only appropriate that I also acknowledge my debt to the many colleagues who have helped me to better understand the complex process of teaching chemistry.

I wish to express my deepest appreciation to those whose encouragement, assistance, and unfailing support have been an essential component in the completion of all three editions of this book, my parents, my children, and most especially, my wife, Virginia.

PREFACE

This study guide/workbook is intended to serve as a supplement to the third edition of *Chemistry and Chemical Reactivity* by John C. Kotz and Keith F. Purcell. The organization is basically the same as in previous editions of the guide; each chapter coincides with a chapter in the textbook. Thus, when a student is working on Compounds and Molecules, Chapter 3 in *Chemistry and Chemical Reactivity*, he or she should consult Chapter 3 of this book. Since the study guide supplements the textbook and class notes, studying will be most effective if all three are used in combination.

The organization of this book has been changed somewhat from previous editions. The listing of new terms has been eliminated, but the Concept Tests have been somewhat lengthened to compensate for this. Two new sections have been added, a Chapter Overview, that attempts to put the current material in context, and Armchair Exercises, which are intended to help the student relate the material to his or her everyday experiences. The learning goals have been continued and are keyed to the individual problems, as was the case in the past. In addition, most chapters include Study Hints that identify some of the more common student misunderstandings and mistakes.

Self testing is widely recognized to be an essential part of any study program. It both reviews what is already known as well as identifies topics that require further work. This guide is organized to encourage a systematic approach to self testing. Each chapter includes a Concept Test, which is intended to evaluate the student's knowledge of the definitions and simple concepts from the chapter. Since these concepts are the building blocks for more complicated material, they should be mastered first.

This study guide provides two further types of self testing. Practice Problems, complete with worked solutions, should provide the student with a good preliminary measure of his or her skills. The practice problems also refer to specific learning goals, so that students spend more time on the topics that seem to be causing difficulties. Finally, the Practice Test gives the student a chance to work under conditions that are as close as possible to those of a real examination.

One of the more helpful components of this study guide should be the chapter entitled "Some Suggestions for the Student." This consists of practical suggestions that should make studying more effective and improve the student's performance in the course. Students are urged to read this section carefully.

This study guide also includes two chapters that do not have counterpart sections in the textbook by Kotz and Purcell. These are labeled "Special Sections" and deal with the use of equivalents and normality for titration calculations. Instructors who follow this approach in their classes should refer their students to this material.

SOME SUGGESTIONS
FOR THE STUDENT

You have probably purchased this book because you are a little worried about taking General Chemistry or perhaps have already gone far enough in the course to recognize that you're having problems. Don't be discouraged; even though you may have heard a great deal about how hard chemistry is, you can still not only pass but even obtain a good grade if you approach the course systematically. This guide is intended to help you organize your work.

Many students find chemistry to be difficult because it requires the mastery of a number of different skills. You must be able to take good notes, organize your work, do mathematical calculations, develop good study habits, do some memorization, and most important, be able to translate problems from word statements into mathematical equations. This isn't as complicated as it sounds, but it does take some practice. Some students who have a good high school background may be able to get by for a while on what they already know, but eventually each individual will have to develop these skills in order to be successful. This chapter contains some suggestions that should help you to develop the skills needed to do well in this course.

There are probably more recommendations in this chapter than any student will wish to apply all at once, so pick a few now, then come back later for more ideas. In order to make this section as useful as possible, the ideas are arranged by topic with individual suggestions indicated by a check mark. Thus if you find that you're having trouble with taking tests, you may wish to go directly to that section and look for suggestions that appear to be useful. The important thing is to come back to this material several times in the coming semester and look for new ways to improve your skills.

LECTURE PREPARATION AND TAKING NOTES

✓ *Read about the material in your textbook before hearing it explained in lecture.*

Don't expect to understand everything in the chapter completely but just try to become familiar with some of the ideas and identify what topics seem hardest to grasp. When you go to the lecture, pay very close attention when these more difficult topics are discussed. If you have had high school chemistry, some of the material in the text will seem familiar. Don't let this give you a false sense of security. Your instructor in college may want some material done differently from the way you learned in high school. He or she has a good reason for wanting this, and if you try to continue to do things the old way, you may find yourself in trouble.

✓ *Read the chapter overview in the study guide.*

This is not a summary of the chapter but instead should help you to recognize how this material fits in with what you have already learned.

✓ *Make your notes as accurate and complete as possible.*

As you listen to the lecture, don't try to write down every word, but instead attempt to determine what the instructor thinks is important and make sure that your notes indicate these priorities.

If the instructor is describing a multi-step problem, he or she may well indicate the proper order in which you should do each step of the procedure. This is often an important part of the problem-solving process, and so your notes should record that order. It may help to circle the starting point, then add arrows to show the progression of from point to point. This becomes very important later in the course, when the problems can be rather complex.

✓ *Review and expand your notes after class is over.*
As soon as you can after each class, review the notes. Reading what you wrote down in class will usually remind you of words and phrases that you didn't have time to include. Add this material to your notes. Be sure to go over your notes as soon as you can, for your memory of these extra comments will soon fade.

Try writing notes only on the right hand page in your notebook, leaving the left page empty. As you read the notes, identify those sections that aren't completely clear. Then try to write a question that focuses on each idea you didn't understand and write it on the left-hand page next to the confusing section. When you have a chance, ask your lecturer or teaching assistant these questions or find the answer in the textbook. Add the answers to the questions in the appropriate place on the left-hand pages.

Once you have a good set of notes, it is time to really read the textbook carefully. If you have trouble with a certain problem type, copy an example problem from your text or study guide into the space on the left-hand page of your notebook. Also add any comments or special data that may be useful. In this way you will create a set of notes that includes supplemental information just where you need it most. Some students rewrite their notes and produce a beautiful notebook. The procedure suggested here may not produce a set of notes that looks as elegant as rewritten notes, but you will probably invest less time, and your notes will be a more clear and complete study aid.

✓ Now try the Armchair Exercises.
Most of the armchair exercises are thought problems; you don't have to physically do them but just think about the results provided. Some of these exercises can be done with simple materials, but remember that you should always exercise caution. Even though every effort has been made to avoid risk, no

experiment is totally without danger. **DO NOT ATTEMPT TO CREATE YOUR OWN CHEMICAL EXPERIMENTS UNLESS YOU ARE SUPERVISED BY AN EXPERIENCED INSTRUCTOR.**

✓ *Don't be afraid to ask questions.*

When you can't understand a topic on your own, ask your instructor or teaching assistant for help. Students rapidly learn that the questions they fail to answer when studying often appear on examinations. Try to make sure that your questions are clearly stated and asked at an appropriate time. Whenever possible, think carefully about your question and make it as specific as possible. Vague questions usually produce answers that aren't very useful.

✓ *When studying, emphasize the material you don't understand.*

As you continue to study the material in a chapter, you should place less emphasis upon the topics that have already been mastered. Unfortunately it's a very human temptation to spend most of your time on the material that you already know. Avoid this by first identifying the topics that need the most work, then making a study plan that will focus on these topics.

✓ *Identify your weaknesses with self-testing.*

The study guide has been designed to encourage self-testing, and you can also write your own tests using worked examples from the book. Try to duplicate the conditions you will encounter on a real examination. Put away all your books and notes, then try to work the test without any help. This will not only reinforce what you do know but also enable you to identify the material that needs more work.

✓ *Consider studying with other students.*

Many students find that it's helpful to study at least part of the time with one or two friends. Sometimes another student understands your problems better than an instructor, and often you can solidify your knowledge by explaining a concept to another student. This type of exchange can be of benefit to everyone. Of course, if you attempt to study with a group and find that all of the time is spent in socializing, that is not a good group for you to join.

USING THE CALCULATOR

You will probably be using a personal electronic calculator in this course. The modern calculator is not only very helpful but also very rugged. There are, however, some steps you can take to make sure that your calculator gives you the best possible service.

✓ *Really learn how to use your calculator.*
Practice using the calculator as much as possible before the first examination. Read the instruction book and try to learn how to use special keys like 1/x and the parentheses that can be extremely useful.

✓ *Take good care of your calculator.*
Although the electronic calculator is quite rugged, it still should be treated with some care. Don't pile books on top of it in your back pack or drop it on the floor. Keep it dry and cool. If you're not sure of how much longer the batteries will last, be sure to have spares when taking an examination.

HOW TO USE THIS STUDY GUIDE

Each component in the Guide has been written with the idea that the best way to study chemistry is to test yourself frequently under the same conditions you will face on the examinations. Put away your textbook and notes and answer questions as though you were working on a test. Self-testing is the best way to reinforce what you already know as well as a good way to discover what topics need more study.

✓ *Make sure that you understand the concepts and terms.*
Once you have studied your notes and the chapter in the text, review the Learning Goals

and do the Concept Test in the study guide. Do you know all of the terms? Do you think you have mastered each skill described in the learning goals? The concept test consists of simple questions on the basic ideas presented in the chapter. Although this section doesn't include all the terms defined in the chapter, if you have trouble with the concept test you probably need to do more studying. If you have trouble with these basics you will probably also have difficulty solving the problems.

✓ *Now you are ready to attempt the practice problems.*

In each case the solution is provided, but don't use it unless you are completely unable to do the problem on your own. Each practice problem is keyed to one of the learning goals. For example, a boldface **L3** after a problem number indicates that it is based on learning goal 3 in the current chapter. The list of learning goals indicates what section of the textbook chapter you should review. At this time you may also wish to review the Study Hints to check for common mistakes that you may be making.

✓ *Next attempt the practice test.*

When you feel you have mastered all of the learning goals, proceed to the practice test. These answers are also related to the learning goals so that you can identify weak areas that you have not yet corrected. Work the practice test on a separate piece of paper, so that you can use the questions later when reviewing for the final.

IMPROVING PROBLEM-SOLVING SKILLS

A major activity in any general chemistry course is solving numerical problems. Since many students have difficulty with this skill, it is an important skill for students to master.

✓ *Try to understand the basic concepts <u>before</u> working the problems.*

If you don't really understand the basic concepts, it's tempting to just memorize the procedures in some of the worked examples and then try to apply what you've memorized like recipes from a cookbook. You may be able to work simple problems using this approach, but there are too many problem types and variations for you to memorize all of the recipes that you will need.

✓ Practice identifying the principle that is the basis of the problem.

Recognizing the key principle which is the basis of a problem is the first and probably the most important step in problem solving. Look for the key words or phrases in the problem that relate to the relationship that is being used. One useful way to practice the recognition of the possible problem types is to use a copying machine to make a copy of all the practice problems at the end of the chapter. Cut out each individual problem, and mix up the resulting scraps of paper. Then try to organize these problems by type, assigning each one to the appropriate learning goal from the study guide. Where it's appropriate, be sure that you know the mathematical relationship or formula required for each learning goal.

Worked examples in the text or guide can be very helpful. Cover the solution and try to set up how you would solve the problem. If you find you are completely stuck, uncover part of the solution for a hint. This kind of practice is especially important when you are studying for the final exam and trying to organize many different problem types.

✓ *Use a systematic approach to problem identification.*

There has been a great deal of research on problem solving techniques, and your instructor may well have some suggestions to help you. The method described in this section is based in part on an excellent book by G. Polya[*] and is suggested as a model of how to attack problems.

First, read the problem carefully and make sure that you understand the question. Clearly identify what data is provided and especially the quantity which you are asked to calculate. Time is important on an examination, but you can waste time by misunderstanding the question and starting to work the wrong problem. Try to organize the information to determine what is important and what is unnecessary. Some students circle the important points in the problem statement or make a rough drawing to help visualize the situation better. A data table is another good way to organize the information, and you will find this method used frequently in the text and study guide, especially when the problems become more complicated.

Second, when you are sure that you understand the problem, develop a plan for solving it. Try to relate this situation to what you already know. Look carefully at the data and the unknown; have you worked a similar problem before? In many cases, a problem that looks difficult is a combination of several easy problems, so try to break it down into simpler components. If you don't have enough information to solve directly for the unknown, can you combine what is available to obtain the missing data? Try to reason backwards; what data would you require to determine the requested unknown? Can you rearrange the available data to produce what you need? If you are totally unable to see how to work a problem, skip it for the time being. Perhaps coming back to the problem later will give you a fresh perspective.

Third, once you have devised a plan, execute it carefully. Don't try to skip steps or do too much in your head. This not only

---

[*] Polya, G. <u>How to Solve It (2nd Edition)</u>, 1971, Princeton University Press, Princeton, NJ

increases the chance of error but also decreases the possibility that you may see your mistake when you check the problem. Students frequently complain that they make "stupid mistakes." When your solution is presented in a clear, step-by-step way, it is easier to avoid simple mistakes as well as to find your errors.

Fourth, check your work if time permits, but even if you are rushed, always take a few seconds to ask yourself if your answer seems reasonable for this type of question. Remember that the instructor may sometimes ask questions that have answers beyond your previous experience, so that every answer that looks unusual isn't necessarily wrong, but often the answer that looks strange reflects a mistake in your solution.

TAKING EXAMINATIONS

✓ *Be sure to get enough sleep the night before the exam.*

Being rested and knowing that you are well prepared should minimize the possibility of text anxiety during the examination. If you are already tired when you start the exam, you won't be able to think clearly and will be more likely to panic. The all-night study session is a well-established college tradition, but it has hurt far more students than it has helped.

✓ *Work carefully and systematically.*

Don't forget to use a systematic approach to problem solving like that described above. Try to distribute your time so that you start with the easier problems. If you become stuck on a hard problem, don't waste all of your time on it. Go on to other questions and come back to the difficult one later if possible. To decrease tension, stop from time to time, look away from your paper, and relax a few seconds. Be sure to use the right number of significant figures and include units with your calculations.

✓ *Learn from your mistakes.*

When your graded test is returned, don't just look at the grade and celebrate (or curse). Go over each question that you missed and try to understand what you did wrong so that you may be able to avoid that error on future examinations. If you find that you are making certain types of mistakes often, then you should check for that error when working tests and quizzes. Often the final exam is cumulative in chemistry courses. If that is the case in your course, you may see similar problems again.

SOME LAST COMMENTS

Even if you do everything suggested, you may not find Chemistry to be an easy subject; many people don't. The goal is to make it less difficult and easier to understand. If this study guide helps to achieve this, it will have accomplished its purpose. Good luck!

CHAPTER 1
MATTER AND MEASUREMENT

CHAPTER OVERVIEW

A main focus of the study of chemistry is the formation of compounds from the chemical elements. This means that you should be able to distinguish chemical changes, which produce new compounds, from physical changes, which do not form new compounds. It also means you should recognize the difference between pure substances, like elements or compounds, and mixtures, which don't have a definite composition and distinct properties.

Quantitative measurements are another essential feature of chemistry, and this chapter discusses many of the basic measurement units which will be used later. Although most temperature readings are given in Celsius degrees, you will also sometimes use Kelvin temperatures or even Fahrenheit temperatures. The conversion from Celsius to kelvin values is especially valuable.

The power of the metric system is that the units are a combination of a base unit with a prefix, which modifies the base unit. The three base units are the meter, the unit of length, the liter, the unit of volume, and gram, the unit of mass. Each prefix has a definite meaning, so if you learn that milli means one thousandth, you now know the size of a milliliter, a millimeter, and a milligram. Comparing the prefixes makes it easy to obtain the conversion factor to convert from one unit to another in the same type of measurement. It's very important to learn the prefixes.

The metric system is also designed to make it easier to convert from one type of measurement to another. For example, cubic length measurements are really volume measurements, and so should be related in a simple way to the standard volume units. This is not the case in English units, since the only way to convert from cubic feet to gallons is to look up the conversion factor in a table. With metric units, the conversion is easier, because one cubic centimeter is equivalent to one milliliter. This means that if you have the dimensions of an object in metric length values, it's easy to convert the lengths into centimeters, then shift to milliliters. Since we use water so often, it's also convenient that one milliliter of water weighs one gram. Even though this value is only true at 4°C, it is often used as an approximation for this conversion.

In order to make the relationships between units more consistent, the metric system has been somewhat modified to create what is called the SI system. This has made some of the familiar units, like the liter, less frequently used, even though the liter continues to be the basis for the names of the units.

Students are sometimes surprised that chemists seem so concerned with the difference between precision and accuracy. The reason is simple. If you are target shooting and always shoot high and to the left, it's relatively easy to adjust the sight so that the shots are now at the center of the target. If the shots are scattered randomly about the target, there is no simple way to improve the accuracy. In the same way, if a chemical analysis has a high level of precision but isn't accurate, it's often easy to identify the error and correct for it.

The widespread use of hand-held calculators has allowed students to calculate answers to many decimal places. You must understand that usually the uncertainty in the numbers used in the calculation is so high that most of the calculator result is worthless garbage. Use the rules for significant figures to make sure that the answer you report is justified by the data used in the calculation.

Percent is a rather simple idea that is often referred to in general chemistry courses. You probably remember it from high school, but make sure that it's perfectly clear in your mind, so that you can use it to solve problems.

LEARNING GOALS

1. Recognize the properties of the gas, liquid, and

solid states of matter; distinguish between chemical and physical properties, and be familiar with the concepts of density and temperature. (Sec. 1.1)

2. Be able to convert between any of the three common scientific temperature scales, Celsius, Fahrenheit, and Kelvin. (Sec. 1.1)

3. Understand the relationships among atoms and elements, molecules and compounds, including knowing how to write elemental symbols and chemical formulas. (Sec. 1.2 and 1.3)

4. Be able to understand the difference between chemical and physical changes. Know the difference between pure substances and mixtures and be able to describe the difference between homogeneous and heterogeneous mixtures. (Sec. 1.4 and 1.5)

5. Convert lengths, masses, and liquid volumes from one SI (metric) unit to another. Also remember the important relationships that 1.000 Liter is defined as 1000 cm^3 and that one milliliter of water is usually considered to have a mass of one gram. The latter relationship is only exactly true at about 4°C but this value is often assumed in the absence of more precise information about the temperature. (Sec. 1.7)

6. It's important to recognize the approximate relationship between the commonly used non-metric units and metric (SI) units, since this type of numerical conversion may be required. Also be able to express large and small quantities in correct scientific notation. (Sec. 1.7)

7. Understand the difference between precision and accuracy and know the guidelines in the textbook for the use of significant figures. Especially notice the difference between the rule for addition and subtraction and that for multiplication and division. Be sure to use dimensional analysis when solving problems. (Sec. 1.7)

8. Review the idea of percentage and be able to use this concept when doing calculations. (1.7)

ARMCHAIR EXERCISES
(Helpful hints are at the end of the chapter.)

1. The textbook refers to the submicroscopic nature of matter. Do you understand what this means? Suppose that you had a magnifying glass powerful enough to see the smallest particles of matter and used it to look at a sample of water. Draw a picture of what you would see.

2. Density is a common concept but still one that can cause confusion. Suppose that a toy boat is floating in a bath tub, and the boat contains a large rock. If the rock is taken out of the boat and placed in the water in the tub, what will happen to the water level? Will it raise, go down, or remain unchanged?

CONCEPT TEST
(See answers at the end of the chapter.)

1. Properties that can be observed and measured without changing the composition of a substance are called _____.

2. The kinetic molecular theory states that all matter consists of tiny particles (atoms and molecules), which are in constant _____.

3. _____ is the property of matter that determines whether heat can be transferred from one body to another and the direction of the transfer.

4. The boiling point of water is _____ degrees on the Celsius temperature scale and is _____ degrees on the Fahrenheit scale.

5. The lowest possible temperature on the Kelvin scale is called _____.

6. A(n) _____ is the smallest particle of an element that displays the chemical properties of the element.

7. _____ have a definite percentage composition (by mass) of their combining elements.

8. When one or more substances are transformed into one or more different substances, this is called a(n) _____.

9. A mixture that is completely uniform at the microscopic level and consists of two or more substances in the same phase is called a(n) _____ mixture.

10. Complete the table below by listing the numerical value represented by each metric prefix:

a. milli _____   b. centi _____

c. kilo _____   d. deci _____

e. micro _____   f. pico _____

11. What is the mass in kilograms of 1.00 liter of pure water? _____.

12. What is the volume in cubic centimeters of a 1.00 liter volume? _____

13. Make the conversions indicated.

a. 2.05 L = _____ mL

b. 0.12 km = _____ m

c. 44.5 cm = _____ mm

d. 450. mg = _____ g

e. 22,500 cm^3 = _____ L

f. 235 mL of water weighs _____ g

14. Indicate the correct number of significant figures in the answer to each calculation.

a. 25.60/2.1 =   b. 5670 x 3.51 =

c. $\dfrac{300. + 56.8}{5}$ =   d. 2.01 + 0.03582 =

15. Round each of the following numbers to two significant figures and report the number in correct scientific notation.

a. 273.15 _____   b. 345,000,000 _____

c. 0.00735 _____   d. 0.00055

STUDY HINTS:

1. This chapter introduces many terms that you must know in order to understand topics later in the course.   Some of the terms may be familiar from your high school chemistry course but don't be fooled by the fact that the term just sounds familiar. A little extra work invested now to make sure that you really understand these terms will prevent confusion later in the course.

2. As in the textbook, the problems in this study guide are worked with an electronic calculator, and the answer is rounded to the correct number of significant figures only at the end of the problem. Rounding off intermediate values may produce slightly different results from those in the text.

3. Students sometimes underestimate the importance of density. It is a simple concept, but one that will be important for many problems later in the course. Be sure to remember it.

4. Many students seem to have difficulty spelling the names of some specific elements. For example, the following names often cause problems (commonly confused letters are underlined): beryllium, fluorine, silicon (no final e), phosphorus, sulfur, chlorine, chromium, nickel, and zinc. Also, don't confuse the symbols of the elements having names that begin with S (sodium, sulfur, silicon, and scandium) or those that begin with P (potassium and phosphorus).

5. Most of the unit conversions you will do in this course will involve only metric units. Remember the prefix indicates the magnitude of the conversion factor, and dimensional analysis will help you to decide whether to divide or multiply by the conversion factor.

6. Remember that all numerical answers should be obtained by carrying through all of the digits on the calculator until the final step and then rounding to the proper number of significant figures. Failure to do this may cause your answer to be significantly different from those in the textbook and study guide.

7. When working with percentage, the first step is usually to convert the percent into a decimal. Don't forget the need for this conversion.

PRACTICE PROBLEMS

1. (L4) Identify each of the following as either a chemical or physical change.

antifreeze boils _____ sugar dissolves _____

gasoline burns _____ a nail rusts _____

2. (L3) Write the correct name for the element represented by each symbol:

a. Si _____ b. Ti _____

c. Mg _____ d. S _____

e. O _____ f. B _____

3. (L5) Make the following temperature conversions.

°F	°C	K
100.	_____	_____
_____	-33	_____
_____	_____	373

4. (L2) A lead block has a length of 0.22 meters, a width of 365 millimeters, a height of 4.8 decimeters, and a mass of 439 kilograms. Based on this information, what is the density of lead in grams/cubic centimeter?

5. **(L4)** If the density of alcohol is 0.789 g/mL, how many mL of alcohol must be measured out for an experiment that requires 2.50×10^2 kilograms of alcohol?

6. **(L3)** The fuel efficiency of a certain automobile is rated as 28.6 miles/gallon. Convert this value into kilometers/liter. (Hint: 1 kilometer = 0.62137 mile and 1 Liter = 1.056710 quarts)

7. **(L2)** The diameter of a single gold atom is 288 picometers. How many gold atoms would one have to set side by side in order to have a line of gold atoms one kilometer long?

8. **(L8)** Yellow brass consists of the elements copper and zinc. If a 555 kilogram sample of yellow brass is 67% copper, how many grams of zinc are present in this sample?

9. **(L4)** The annual production of sulfuric acid in the United States varies, but it is approximately 30,000,000 tons per year. If the density of concentrated sulfuric acid is 1.85 grams/mL, how many liters of concentrated sulfuric acid are produced each year? (Hint: 1 pound = 453.6 grams)

PRACTICE PROBLEM SOLUTIONS

1. antifreeze boils - physical
   sugar dissolves   - physical
   gasoline burns   - chemical
   a nail rusts   - chemical

2. a. silicon   b. titanium
   c. magnesium   d. sulfur
   e. oxygen   f. boron

3.

°F	°C	K
100.	37.8	311
−27	−33	240
212	1.0×10^2	373

4. First find the volume in cm^3.

$$Volume = length \times width \times height$$

Volume =
  0.22 m x 100 cm/m x 365 mm
                  x 1 cm/10 mm x 4.8 dm x 10 cm/dm

$$V = 38544 \ cm^3$$

Now convert the mass to grams and substitute in the density equation.

$$Density = mass/volume$$

$$D = \frac{439 \ kg \times 1000 \ g/kg}{38544 \ cm^3}$$

$$\underline{D = 11 \ g/cm^3}$$

5. This problem can be solved simply by substituting in the density equation.

$$D = M/V$$

rearrange the equation, insert the mass and density values given as well as the conversion factors for kg to g and mL to liters.

$$V = M/D \ = \frac{2.50 \times 10^2 \ kg \times 1000 \ g/kg}{0.789 \ g/mL}$$

$$\underline{V = 3.17 \times 10^5 \ mL}$$

6. If you remember the conversion factors, this is a simple problem for dimensional analysis.

Mileage Rating

$= \dfrac{28.6 \ miles}{gal} \times \dfrac{1 \ km}{0.62137 \ mile} \times \dfrac{1 \ gal}{4 \ qts} \times \dfrac{1.056710 \ qt}{1 \ Liter}$

$$\underline{Mileage \ Rating = 12.2 \ km/L}$$

7. Dimensional analysis provides the key to this problem also.

Total length =
       number of gold atoms x length of one atom

Rearrange to isolate the unknown

number of gold atoms

$$= \frac{1 \text{ kilometer}}{\dfrac{288 \text{ pm}}{1 \text{ Au atom}} \times \dfrac{1 \text{ m}}{1 \times 10^{12} \text{ pm}} \times \dfrac{1 \text{ km}}{1000 \text{ m}}}$$

number of gold atoms = 3.47×10^{12} atoms

8. Since the brass consists of only two elements, the percentage of zinc can be found by subtracting

       percent zinc = 100 - percent copper
                    = 100 - 67
                    = 33%

Convert this percent to a decimal and multiply times the total mass

       mass of zinc = 0.33 x 555 kilograms

       mass of zinc = 180 kilograms

convert to grams
       mass of zinc = 180 kg x 1000 g/kg

       mass of zinc = 1.8×10^{5} grams

9. Again, this problem is best done with dimensional analysis.

grams of H_2SO_4

   = 3×10^{7} tons x 2000 lb/ton x 453.6 g/lb

grams of H_2SO_4 = 2.7216×10^{13} g

$$\text{volume of } H_2SO_4 = \frac{2.7216 \times 10^{13} \text{ g}}{1.85 \text{ g/mL} \times 1000 \text{ mL/L}}$$

<u>volume of H_2SO_4 = 1×10^{10} liters</u>

PRACTICE TEST

Put away all of your books and notes and work on this test as though it were an actual examination. You may use a periodic table, if necessary. Allow yourself 30 minutes to do the test. The answers are at the end of this chapter.

1. Identify each of the following as either a chemical or physical change.

ice melts _____ silver tarnishes _____

steam condenses_____ perfume evaporates _____

2. Which of the element names below is spelled incorrectly? (Note, there may be more than one.)

a. phosphorous   b. florine

c. beryllium   d. silicone

e. vanadium   f. clorine

3. The temperature at the surface of the planet Venus is about 470°C. Convert this temperature to (a) degrees Fahrenheit and (b) kelvins.

4. Make the following conversions as indicated.

a. 45.0 mm = _____ cm b. 0.3 km = _____ cm

c. 3.9 kg of water has a volume of _____ liters.

5. Suppose that you wish to purchase a water bed that has the dimensions 2.45 m x 21.5 dm x 23 cm. How many kilograms of water does this bed contain?

6. A certain bronze sample consists of 22% tin and 78% copper. How many grams of tin are present in a 1.2 kilogram sample of this type of bronze?

12

7. The density of platinum metal is 21.5 g/cm^3. Calculate the mass of a block of platinum that has the dimensions 81.0 cm x 37 mm x 0.023 m.

8. The solubility of salt, NaCl, in water is 360 gram/liter. How many kilograms of salt will dissolve in 0.52 gallons of water?

9. Suppose that someone offers you a golden opportunity to invest in an industrial plant that will extract gold from sea water. The only information available about the plant is that it is expected to process 100,000,000 kilograms of sea water a day. From a chemical handbook you learn that the density of sea water is 1.03 g/mL and sea water contains 4.0×10^{-5} mg/L of gold. What is the maximum number of grams of gold that the plant could possibly produce per day?

CONCEPT TEST ANSWERS

1. physical properties   2. motion
3. temperature   4. 100 deg, 212 deg
5. absolute zero   6. atom
7. compounds (or molecules)
8. chemical change or chemical reaction
9. homogeneous
10. a. 1/1000   b. 1/100   c. 1000
     d. 1/10   e. 1×10^{-6} f. 1×10^{-12}
11. 1.00 kilograms

12. 1.00×10^3 cubic centimeters

13. a. 2.05×10^3 mL   b. 1.2×10^2 m
    c. 445 mm   d. 0.450
    e. 22.5 L   f. 235 g
14. a. 2   b. 3   c. 1   d. 3
15. a. 2.7×10^2 b. 3.5×10^8
    c. 7.4×10^{-3} d. 5.5×10^{-4}

PRACTICE TEST ANSWERS

1. (L4) ice melts   – physical
     silver tarnishes   – chemical
     steam condenses   – physical
     perfume evaporates – physical

2. **(L3)** phosphorus, fluorine, silicon, chlorine
3. **(L2)** a. 880 $^{\circ}$F   b. 740 K
4. **(L5)** a. 4.50 cm   b. 3×10^4 cm
           c. 3.9 liters
5. **(L5)** 1.2×10^3 kg   6. **(L8)** 260 grams
7. **(L1 and L5)** 1.5×10^4 grams
8. **(L1 & L5)** 0.71 kilograms of NaCl
9. **(L5)** 4 grams of gold per day

CHAPTER 2
ELEMENTS AND ATOMS

CHAPTER OVERVIEW

Like Chapter 1, this chapter develops some of the basic language and concepts of chemistry. Building a good foundation here will pay dividends as you proceed to more difficult material.

The idea that matter consists of tiny particles called atoms is probably already familiar to you, even though Dalton's actual assumptions may not be as well-known. The law of definite proportions is important because it is the basis for our identification of compounds; a specific compound must always have the same composition. Don't overlook the law of conservation of mass. There are a number of problem types that are easy if you recognize that they are based on this law.

The nuclear model of the atom, as first proposed by Rutherford, considers three fundamental particles, protons, neutrons, and electrons, which are the building blocks of atoms. Simply remember that the atomic number (Z) determines which element is being considered and it equals the number of protons. The mass number (A) equals sum of the protons and neutrons. For neutral atoms, the number of electrons must equal the number of protons, but for ions, the number of protons remains fixed for a given element, and the number of electrons varies to produce the observed charge.

Most of the elements as observed in nature consist of several possible isotopes. For a given element all the isotopes have the same atomic number but have different atomic masses. Since the atomic number can't change for a given element, this variation must be caused by different numbers of neutrons.

It's possible to predict the properties of elements based on where they are found on the periodic table. As the course continues, the position of an element will be used to predict not only its properties, such as conductivity and malleability, but also to predict reactivity.

Learning to identify the groups now prepares you for this type of work.

The final, and perhaps the most important topic in this chapter is the use of Avogadro's number to convert from grams to moles. As noted in the study hints section, this is an essential calculation, which you will perform many, many times before the semester is over.

LEARNING GOALS

1. Understand the basic assumptions of Dalton's Atomic Theory of Matter, including his hypothesis that is now known as the law of conservation of matter. You should also know the law of definite proportions and the law of multiple proportions. (Sec. 2.1)

2. Be able to explain Rutherford's nuclear model of the atom, including the three primary constituents of the atom, namely, electrons, protons, and neutrons. Given the atomic number, Z, and mass number, A, for an element having the symbol X, be able to represent that atom using the notation

$$_{Z}^{A}X$$

Also be familiar with the relationships involving atomic number, mass number, number of protons, number of neutrons, and number of electrons. For ions, be able to determine the number of electrons from the charge or vice versa, and also to represent the ion using a form like that shown above. (Secs. 2.2 and 2.3)

3. Be able to calculate the average atomic mass of an element from the relative abundances of the component isotopes and also calculate the isotopic abundances for elements where only two isotopes of known masses exist. (Secs. 2.4 & 2.5)

4. Be familiar with the periodic table and be able to locate the alkali metals, the alkaline earth metals, the transition metals, groups 4A and 5A, the chalcogens, the halogens, and the noble gases on the table. Using the periodic table, be able to

identify an element as a metal, a nonmetal, or a metalloid and also identify main group and transition elements. (Sec. 2.6)

5. Since grams, moles, and number of atoms (or molecules) are closely related, be able to convert from one of these to another for both elements and compounds. In order to do this it will also be necessary to understand and use atomic weights and molecular (or formula) masses. (Sec. 2.7)

ARMCHAIR EXERCISE
(Helpful hints are at the end of the chapter.)
1. Avogadro's number is so large that it can be difficult to fully understand its size. The following thought exercise may be helpful in understanding how large this number is.

Imagine that you have a pair of tweezers that would enable you to pick up exactly 1000 molecules at a time. If you use these tweezers to remove 1000 molecules per second from a beaker that contained 18.0 mL, that is, one mole, of water, and if no evaporation occurred, how long would it take to remove all of the water from the beaker?

Think about the situation before you start calculating. Would there be any water left after removing 1000 water molecules per second for a year? Would there be any left after a century? What about a million years? Once you have made your guess, calculate and see how close you came.

2. Everyone knows that ice floats in liquid water. Suppose that you observed a colorless liquid, which looked like water, and what appeared to be ice cubes were present, but the cubes had sunk to the bottom of the liquid. How might you explain this?

CONCEPT TEST
(Answers are provided at the end of the chapter.)
1. Dalton's Theory states that all matter is composed of indivisible and indestructible particles called _____.

2. Lavoisier proposed the law of conservation of matter, which states that there is no change in

_____ when a chemical reaction occurs.

3. When a particular compound is purified, it always contains the same _____ in the same ratio by mass.

4. Benjamin Franklin proposed a fundamental rule of electricity, that like charges attract each other, but unlike charges _____.

5. Passing an electric current through a solution of a compound to cause a chemical reaction is called _____.

6. Chadwick discovered the fundamental particle called the _____, which has no electric charge and has almost the same mass as the proton.

7. Because atoms normally have no charge, the number of _____ and _____ in an atom must be equal.

8. The _____, or core, of the atom is made up of protons and neutrons.

9. The number of protons in the nucleus of atoms of a given element is called the _____.

10. The atomic masses of the elements are measured relative to the mass of an atom of the element _____ that has six protons and six neutrons.

11. The _____ of a particular atom, indicated by the symbol A, indicates the total number of protons and neutrons in the atom.

12. _____ are atoms having the same atomic number but different mass numbers.

13. Elements may be classified as metals, nonmetals, or metalloids. Which of these types of

element is most likely to conduct an electric current? _____ Which type is most likely to be a gas? _____

14. The modern statement of the law of chemical periodicity is that the properties of the elements are periodic functions of _____.

15. The mass in grams of one mole of atoms of any element is the _____.

16. _____ is the number of particles in one mole of any substance.

17. To find the molar mass of the compound Fe_2O_3, add together the mass of _____ moles of iron atoms and _____ moles of oxygen atoms.

18. Identify each element below as a metal, a nonmetal, or a metalloid.

a. Si _____   b. K _____

c. Mn _____   d. S _____

e. Na _____   f. Fe _____

STUDY HINTS

1. The authors of the textbook emphasize the importance of learning to convert from moles to grams and from grams to moles, but it is worth stressing this point again. This simple conversion is a basic step in a great many of the problems encountered in the early part of the course. Learning to do it well now will save much frustration during the coming semester.

2. Students sometimes become confused because a single chemical symbol can have several different meanings. For instance, Fe can mean
    a. one atom of iron,
    b. one mole of iron atoms, or
    c. one molar mass of iron.
When reading chemical symbols, remember that all of

these interpretations are possible. You must decide which one is most appropriate in a given case.

3. The quantity that the textbook calls molar mass may also be referred to as gram molecular mass or gram molecular weight. If these terms are encountered, try to remember they are different names for basically the same quantity.

PRACTICE PROBLEMS

1. (L2) Determine the number of protons, neutrons, and electrons in the following species

$$\text{(a) } {}^{7}_{3}\text{Li} \qquad\qquad \text{(b) } {}^{26}_{12}\text{Mg}^{2+}$$

a. ___ protons, ___ neutrons, and ___ electrons.

b. ___ protons, ___ neutrons, and ___ electrons.

2. (L5) Complete the table.

number	B	Ar	Cu
of grams	1.9	_____	_____
atomic mass	_____	_____	_____
moles	_____	113	_____
number of atoms	_____	_____	2.0×10^{20}

3. (L4) Classify each of these elements as either a metal or a nonmetal.

a. Ni _____   b. N _____

c. Cr _____   d. Li _____

4. (L4) Classify each of the elements in the above question as either a main group or a transition element.

main group _____   transition _____

20

5. (L3) The element neon has three stable isotopes, with masses and abundances as shown below. What is the average atomic mass of neon? (Note: Due to experimental error, the total percent is not 100%)

Exact Mass	Relative Abundance (%)
19.9924	90.84
20.9940	0.259
21.9914	8.90

6. (L5) Element 15 is found in a variety of useful products, including fertilizers, detergents, and even matches. In 1984 approximately 3.3×10^8 kilograms of this element was produced in the United States. What is this element and how many moles of it were produced?

7. (L5) If the average mass is 3.16×10^{-23} grams for a single atom of a certain element, what is the atomic mass of this element?

8. **(L2)** Give the number of protons and electrons for each of the following ions.

no. of protons   no. of electrons

O^{2-}

_____   _____

Li^+

_____   _____

Mg^{2+}

_____   _____

9. **(L4)** Consider the following list of elements, chlorine, iron, and sodium. Which one(s) would you expect to be most likely to display each of the characteristics listed below:

a. Which would most likely be a gas?

b. Which would most likely have a metallic luster?

c. Which would be least likely to conduct electricity?

PRACTICE PROBLEM SOLUTIONS

1. a. The atomic number is 3, and so the <u>number of protons = 3</u>. Since this is a neutral species, the <u>number of electrons</u> must also equal the atomic number, or <u>3</u>.

The mass number, 7, must equal the sum of the number of protons and neutrons. Since the number of protons has already been found to be 3, then

mass number

= number of protons + number of neutrons

7   =   3   + number of neutrons

<u>number of neutrons = 7 - 3 = 4</u>

1b. The atomic number is 12, so the <u>number of protons = 12.</u>

Since this is a charged species,
ion charge

= number of protons - number of electrons

+2   =   12   -   number of electrons

22

<u>number of electrons = 10</u>

Finally, using the mass number relationship

mass number
      = number of protons + number of neutrons
  26 =   12   + number of neutrons

      <u>number of neutrons = 14</u>

2.   B   Ar   Cu
number
   of grams   **1.9**   <u>4510</u>   <u>0.021</u>

atomic mass <u>10.8</u>   <u>39.95</u>   <u>63.5</u>

moles   <u>0.18</u>   **113** 3.3×10^{-4}

number
of atoms   <u>1.1×10^{23}</u>   <u>6.80×10^{25}</u>   **2.0×10^{20}**

3. metals: Ni, Cr, Li   nonmetals: N

4. typical: N, Li   transition: Ni, Cr

5. atomic mass
      = (isotope 1 abundance)(isotope 1 mass)
       +(isotope 2 abundance)(isotope 2 mass)
       +(isotope 3 abundance)(isotope 3 mass)

atomic mass
      = 0.9084x19.9924 + 0.00259x20.9940
               + 0.0890x21.9914

<u>atomic mass = 20.17 grams/mole</u>

6. From the periodic table, the element having an atomic number of 15 is phosphorus, and the molar mass of phosphorus is 30.9738 grams/mole.

$$\text{mol of P} = \frac{3.3 \times 10^{8} \text{ kg} \times 1000 \text{ g} \times 1 \text{ mole}}{1 \text{ kg} \times 31.0 \text{ g}}$$

<u>mol of P = 1.1×10^{10}</u>

7. Atomic mass =
      3.16×10^{-23} grams/atom x $6.02 \times 10_{23}$ atoms/mole

Atomic mass = 19.0 grams/mole

	no. of protons	no. of electrons
$^{8}_{}O^{2-}$	8	10
Li^{+}	3	2
Mg^{2+}	12	10

10. a. gas: chlorine

b. metallic luster: iron, sodium

c. not conduct electricity: chlorine

PRACTICE TEST
Allow 30 minutes to complete this practice test.

1. The element copper, which has an average atomic weight of 63.54 grams/mole, has two stable isotopes, copper-63 (mass = 62.9298) and copper-65 (mass = 64.9278). What percent of naturally occurring copper is Copper-63?

2. Determine how many protons, neutrons, and electrons are present in

a. $^{18}_{8}O^{2-}$ and b. $^{60}_{27}Co^{3+}$

3. Write the name of the element that matches each symbol and indicate whether each is a metal, metalloid, or nonmetal.

a. Zn   b. Sc   c. O   d. Al

4. Name each of the following elements and identify it as an alkali metal, an alkaline earth metal, a halogen, or a transition metal.

a. F   b. Li   c. Be   d. Ti   e. Ca

5. What is the mass in grams of a single atom of an element that has an atomic mass of 28.1 g/mole?

6. Complete the table.

	Na	Si	Ne
number of grams	_____	67.2	_____
atomic mass	_____	_____	_____
moles	_____	_____	0.025
number of atoms	5.5×10^{21}	_____	_____

7. Give the number of protons and electrons for each of the following ions.

	no. of protons	no. of electrons
Cu^{2+}	_____	_____
B^{3+}	_____	_____
F^{-}	_____	_____

8. When a 2.00 gram sample of mercury(II) oxide is heated, the products are liquid mercury and gaseous oxygen. If 1.85 grams of mercury are produced, how many grams of oxygen must also result? How many moles of mercury are produced? How many mercury atoms are produced?

9. If a sample of iron consists of 5.0×10^{20} iron atoms, how many moles of iron are present in the sample? How many grams of iron are present in the sample?

COMMENTS ON ARMCHAIR EXERCISES

1. Of course, tweezers like those in this question don't really exist, but the results of this exercise should help you to understand how incredibly large Avogadro's number actually is.

2. The obvious answer is that the liquid isn't really water. There are a number of liquids that look very much like water, but have a solid phase that is more dense than the liquid. Another option that you might have considered is that the ice cube

actually was water, but that it was water with the heavy isotopes of hydrogen and/or oxygen.

CONCEPT TEST ANSWERS

1. atoms
2. mass
3. atoms (or elements)
4. repel
5. electrolysis
6. neutron
7. protons, electrons
8. nucleus
9. atomic number
10. carbon
11. mass number
12. isotopes
13. metal, nonmetal
14. atomic number
15. molar mass of the element
16. Avogadro's number(6.022×10^{23})
17. two, three
18. K, Mn, Na, and Fe are metals, S is a nonmetal, and Si is a metalloid.

PRACTICE TEST ANSWERS

1. (L3) 69.46% copper-63
2. (L2) a. 10 neutrons, 8 protons, and 10 electrons
   b. 33 neutrons, 27 protons, and 24 electrons
3. (L4) a. zinc   metal
   b. scandium metal
   c. oxygen   nonmetal
   d. aluminum metal
4. (L4) a. F fluorine   halogen
   b. Li lithium   alkali metal
   c. Be beryllium alkaline earth
   d. Ti titanium   transition metal
   e. Ca calcium   alkaline earth
5. (L5) 4.67×10^{-23} grams
6. (L5)

	Na	Si	Ne
number of grams	2.1×10^{-1}	67.2	5.1×10^{-1}
atomic mass	23.0	28.09	20.2
moles	9.1×10^{-3}	2.39	0.025
number of atoms	5.5×10^{21}	1.44×10^{24}	1.5×10^{22}

7. (L2)

	no. of protons	no. of electrons
Cu^{2+}	29	27
B^{3+}	5	2
F^-	9	10

8. **(L1 and L5)** The reaction produces 0.15 grams of oxygen, 0.00922 moles of mercury, or 5.55×10^{21} mercury atoms.

9. **(L5)** 4.6×10^{-2} grams of iron, 8.3×10^{-4} moles of iron

CHAPTER 3
COMPOUNDS AND MOLECULES

CHAPTER OVERVIEW

This chapter continues to develop the basic language of chemistry. The concepts discussed here are the basis of much that comes later and so continued attention will pay future dividends.

Both compounds and also some elements exist in the form of molecules. Because molecules are such an important part of chemistry, there are several different ways to represent molecules. Empirical formulas are simplest and only show the mole ratios of the elemental components. Molecular formulas show the actual composition of the molecules, and structural formulas show how the individual atoms are connected to each other. Work carefully to make sure you understand these differences now.

Many compounds exist as combinations of ions rather than as molecules. In order to write these formulas, you must be able to identify the cations and anions that are likely to result when elements form compounds, then combine these ions in the correct ratio to create the compound. Some ions are combinations of several elements, which is you must learn the names and formulas of these polyatomic ions in order to understand how they participate in compound formation.

The conversions involving moles and atoms, which were important in the last chapter, are now extended to molecules. As noted earlier, you will do this type of calculation frequently in the days ahead. Working with molecules requires that you be able to determine the molar mass of a compound from its formula. obtain the empirical formula from the percent composition of a compound, and calculate the percent composition from the formula.

Determining the formula for a hydrated compound combines several of the important concepts that are discussed in this chapter, and so is an excellent review of this material.

28

LEARNING GOALS

1. Recognize the elements that exist as diatomic molecules and also the special molecular configurations for elements like sulfur and phosphorous. (Sec. 3.1)

2. Recognize and be able to explain several different ways of representing molecular compounds, including molecular formulas and structural formulas. (Sec. 3.2)

3. Identify the elements most likely to exist as cations and those most likely to exist as anions. Determine the charge for monatomic ions. (Sec. 3.3)

4. Know the names, formulas, and charge of the common polyatomic ions. (Sec. 3.3)

5. Predict the formulas for simple ionic compounds and understand the characteristic properties of this type of compound. Given a simple ionic compound, recognize what component ions are present and give their relative number. (Sec. 3.4)

6. Know how to name and write formulas for ionic compounds, binary compounds of the nonmetals, and selected hydrocarbons using the rules explained in the textbook. (Sec. 3.5)

7. Given the formula of a compound you should be able to determine the molar mass (or formula mass). The relationships that allow conversions among quantities such as grams, moles, number of molecules, and number of atoms of component elements are among the most fundamental in chemistry. You should also be able to convert between empirical formulas and molecular formulas. Be sure that you understand these relationships. (Sec. 3.6)

8. You should be able to calculate percent composition from empirical formula or vice versa and also be able to obtain the molecular formula from the empirical formula. (Sec. 3.7)

9. Be able to apply the concepts of molecular composition outlined in the previous learning goal

to special cases, such as the composition of hydrated compounds. (Sec. 3.8)

ARMCHAIR EXERCISE

1. Can two different compounds have the same empirical formula? If you this is possible, what would be the conditions where this might exist?

2. What do all of the following have in common: chalk, limestone, marble, and some antacid tablets? Is this surprising? Can you explain why it's true?

CONCEPT TEST
(Answers are provided at the end of the chapter.)
1. Circle the symbol or symbols below for the pure elements that are <u>not</u> diatomic.

He   H   I   P   O   Na

2. _____ are different forms of the same element.

3. Organic compounds invariably contain the two elements _____ and _____.

4. A(n) _____ is a modified form of a molecular formula in which the elemental symbols are arranged to emphasize the connectivity of atoms and the important groups of atoms in the molecule.

5. Metal atoms usually lose _____ to form positively charged ions called _____.

6. Nonmetal atoms usually gain _____ to form negatively charged ions called _____.

7. Ions having the same number of electrons as a(n) _____ are especially favored when compounds form.

8. When writing a chemical formula, the symbol of the _____ is always given first, followed by the symbol for the anion.

9. The electrostatic force between oppositely charged ions is described by _____ law.

10. When forming ionic compounds, the elements classed as _____ usually form positive ions.

11. How many moles of oxygen atoms are present in one mole of each of the compounds below?

a. H_2O b. $NiCl_2 \cdot 6H_2O$ c. $HClO_4$

        d. $BaCl_2 \cdot 2H_2O$ e. $Ca(OH)_2$

12. To find the molar mass of the compound Fe_2O_3, you must add together the mass of _____ moles of iron atoms and _____ moles of oxygen atoms.

13. The _____ is the mass in grams of Avogadro's number of molecules.

14. The law of constant composition states that any sample of a pure compound always consists of the same _____ combined in the same _____

_____.

15. The molecular formula is always the product of _____ multiplied times the subscripts in the empirical formula.

16. Name each of these polyatomic ions:

a. NO_3^- _____   b. OH^- _____

c. SO_4^{2-} _____   d. PO_4^{3-} _____

17. Write the formula for each polyatomic ion:

a. ammonium _____   b. sulfate _____

c. acetate _____   d. carbonate _____

18. Circle the hydrated compound or compounds in the following list.

a. H_2O b. $NiCl_2 \cdot 6H_2O$ c. $HClO_4$

        d. $BaCl_2 \cdot 2H_2O$ e. $Ca(OH)_2$

STUDY HINTS

1. Remember that under normal conditions pure hydrogen, nitrogen, oxygen, and the halogens exist as diatomic molecules, whereas the noble gas elements exist as uncombined atoms.

2. Notice that the properties of an ion are quite different from the properties of the atom from which the ions was derived. For example, the element chlorine is a deadly poison gas, but the chloride ion, as found in common table salt, is relatively nontoxic for most people. Thus, it's very important to notice whether you are dealing with an element or an ion derived from the element.

2. When working problems related to chemical formulas, always remember that a formula is basically a mole ratio. Therefore, the initial stage of such a problem often involves the determination of the number of moles of the components in the formula.

3. Like chemical symbols, chemical formulas can have several different meanings. For instance, H_2O can mean

    a. one molecule of water,
    b. two atoms of the element hydrogen combined with one atom of oxygen,
    c. one mole of water molecules, or
    d. one molar mass of water molecules.

Be sure that you consider these possibilities when you read either a chemical symbol or formula.

4. Students often have difficulty with problems involving a mass decrease due to the formation of a gas, such as oxygen or carbon dioxide. Once you learn to recognize this problem type, it really isn't as difficult as it looks. Just remember that the decrease in mass is due to the escape of a gas, so the mass change represents the mass of gas produced. Use this value to find the moles of gas, which is usually the key to solving the problem. Also be alert for a variation of this problem type where the mass increases due to the combination of the original solid with a gas, like oxygen. Again, the key step is to calculate the moles of gas from the mass change.

PRACTICE PROBLEMS

1. **(L7)** How many moles of phosphorus atoms are contained in one mole of each of the following compounds?

a. P_2O_5 _____ b. $Ca_3(PO_4)_2$ _____ c. $Na_5P_3O_{10}$ _____

2. **(L6)** Name each compound

a. MnS_2 _____ b. $KMnO_4$ _____

c. $TiCl_4$ _____ d. NaCN _____

3. **(L6)** Write the formula for each compound

a. ammonium nitrate   _____

b. iron(III) chloride   _____

c. ethane   _____

d. oxygen difluoride   _____

e. vanadium(IV) sulfate _____

f. aluminum trichloride _____

g. propane   _____

4. **(L5)** Write formulas for all of the compounds that can be formed by combining each cation with each of the anions listed. Name the resulting compounds.

CATIONS	ANIONS
Li^+	F^-
Ca^{2+}	PO_4^{3-}

Answers

_____   _____

_____   _____

5. **(L7)** Assume that you have 15 grams of each of the following compounds. How many moles does that represent in each case?

a. NH_3 b. NH_4NO_3 c. NH_2COONH_4

6. (L7) Hydrazine, N_2H_4, was used as a rocket fuel by the Germans in World War II.

a. What is the molar mass of hydrazine? _____

b. How many moles of hydrazine are contained in a pure sample of N_2H_4 having a mass of 9.6 grams? _____

c. How many molecules of N_2H_4 are present in this 9.6 gram sample? _____

d. How many hydrogen atoms are present in a 9.6 gram sample of N_2H_4? _____

7. (L7) The empirical formula of a certain compound is CH_2 and the molar mass is 84.0 g/mole. What is the molecular formula of this compound?

8. (L6) The first true compound of a noble gas was isolated in 1962 and found to have the composition 29.8% xenon, 44.3% platinum, and 25.9% fluorine. What is the empirical formula of this compound?

9. (L7) Natural gypsum is a hydrated form of calcium sulfate having the formula $CaSO_4 \cdot xH_2O$. When 2.00 grams of gypsum is heated at $200°C$ until all of the water has been driven off, the mass of the remaining solid is 1.58 grams. What is the value of x in the formula for gypsum?

PRACTICE PROBLEM SOLUTIONS

1. a. two moles   b. two moles   c. three moles

2. a. manganese(IV) sulfide
   b. potassium permanganate
   c. titanium(IV) chloride
   d. sodium cyanide

3. a. NH_4NO_3 b. $FeCl_3$ c. C_2H_6

   d. OF_2 d. $V(SO_4)_2$ e. $AlCl_3$

   f. C_3H_8

4.   LiF   lithium fluoride

        Li_3PO_4 lithium phosphate

        CaF_2 calcium fluoride

        $Ca_3(PO_4)_2$ calcium phosphate

5. a. NH_3 15 g/17.0 g/mol = 0.88 moles

   b. NH_4NO_3 15 g/80.0 g/mol = 0.19 moles

   c. NH_2COONH_4 15 g/78.0 g/mol = 0.19 moles

6. a. 32 g/mole   b. 3.0×10^{-1} mole

   c. 1.8×10^{23} molecules   d. 7.2×10^{23} atoms

7. The apparent molar mass of CH_2 is 14
To find how many times this unit is contained in
84 g/mole, divide

$$\frac{84}{14} = 6$$

Therefore the molecular formula is 6 times the
empirical formula, or

$$\underline{\quad C_6H_{12} \quad}$$

8. First assume 100.0 grams of compound in order to
obtain masses of the component elements

mole Xe = 29.8 g / 131.3g/mole   = 0.2270 mole Xe

mole Pt = 44.3 g / 195.1 g/mole = 0.2271 mole Pt

mole F   = 25.9 g / 19.0 g/mole   = 1.363   mole F

To obtain the mole ratio, divide each number of moles by 0.2270, the smallest number of moles.

Xe $\dfrac{0.2270}{0.2270} = 1$ Pt $\dfrac{0.2271}{0.2270} = 1$ F $\dfrac{1.363}{0.2270} = 6$

Thus the ratio Xe:Pt:F :: 1:1:6, and the formula is

$$\underline{XePtF_6}$$

9. To find x, it is necessary to calculate the ratio of the moles of water to the moles of nonhydrated $CaSO_4$.

mole H_2O $= \dfrac{2.00g - 1.58\ g}{18.0\ g/mole} = 0.02333$ moles H_2O

mole $CaSO_4 = \dfrac{1.58\ g}{136\ g/mole} = 0.01160$ mole $CaSO_4$

$$\dfrac{\text{mole water}}{\text{mole calcium sulfate}} = \dfrac{0.02333\ \text{moles}}{0.01160\ \text{moles}} = 2$$

$\underline{x = 2}$ and the formula is $CaSO_4 \cdot 2H_2O$

PRACTICE TEST (Allow yourself 35 minutes)

1. How many moles of sulfur atoms are contained in one mole of each of the following compounds?

a. CS_2 b. $Na_2S_2O_4$ c. $Ca(HSO_3)_2$

2. Sodium thiosulfate, which has the formula $Na_2S_2O_3 \cdot 5H_2O$, is used in the photographic and paper-making industries. What is the molar mass of sodium thiosulfate and the percentage of sulfur in this compound.

3. Write the formula for each compound listed.

a. boron trichloride   b. copper(II) chloride
c. aluminum nitrate   d. sodium sulfide
e. methane   f. butane

4. Write the name for each compound listed.

a. KCH_3COO

b. $TiCl_4$

c. CaI_2

d. NH_4ClO_4

5. For each ionic compound, give the formula, charge, and number of each component ion.

a. $(NH_4)_2SO_3$ b. $CaCl_2$ c. K_2SO_4 d. $Mg_3(PO_4)_2$

6. Answer the following questions regarding the compound phosgene, $COCl_2$, a World War I poison gas.

a. What is the molar mass of phosgene. _____

b. How many moles of phosgene are contained in a pure sample of $COCl_2$ having a mass of 2.00 grams? _____

c. How many molecules of $COCl_2$ are present in this 2.00 gram sample? _____

d. How many chlorine atoms are present in a 2.00 gram sample of $COCl_2$? _____

7. Fertilizers are one of the chemical products produced in the greatest quantity. Determine which of the two phosphorus-containing compounds listed will contain the higher percentage of phosphorus.

a. $Ca(H_2PO_4)_2$

b. $NH_4H_2PO_4$

8. Calculate the atomic weight of the unknown element M, if the molar mass of the compound $Na_2M_2O_3$ is 156 g/mol.

9. Mirex is a pesticide that breaks down slowly in the environment but is very useful for controlling certain insect pests. Mirex is composed of 21.9% carbon and 78.1% chlorine, and has a molar mass of 545 g/mol. (a) what is the empirical formula of mirex? (b) What is the molecular formula of mirex?

10. When solid magnesium carbonate is heated, a portion of it decomposes to form solid magnesium oxide and gaseous carbon dioxide. If 2.25 grams of magnesium carbonate is heated until the remaining solid, a mixture of MgO and unreacted $MgCO_3$, has a mass of 1.95 grams, what percentage of the $MgCO_3$ has decomposed?

COMMENTS ON ARMCHAIR EXERCISES

1. Since the empirical formula represents the simplest ratio, it is quite possible for more than one compound to have the same formula. For example, both NO_2 and N_2O_4 have the same empirical formula.

2. All four of these are calcium carbonate. Notice that there can be significant variations in the appearance of the same compound. This is why the determination of the formula is so important.

CONCEPT TEST ANSWERS

1. Helium (He), phosphorus (P), and sodium (Na) are not diatomic.
2. allotropes   3. carbon and hydrogen
4. structural formula   5. electrons, cations
6. electrons, anions   7. noble gas atom
8. cation   9. Coulomb's
10. metals
11. a. one b. six c. four d. two e. two
12. two, three   13. molar mass
14. elements, proportions by mass
15. a small, whole number
16. a. nitrate   b. hydroxide
    c. sulfate   d. phosphate
17. a. NH_4^+ b. SO_4^{2-}

    c. $C_2H_3O_2^-$ d. CO_3^{2-}
18. $NiCl_2 \cdot 6H_2O$ and $BaCl_2 \cdot 2H_2O$ are hydrates.

PRACTICE TEST ANSWERS

1. (L5) Each compound contains two moles of sulfur per mole of compound.
2. (L4 & L6) molar mass = 248.17 g/mol
   The percent of sulfur is 25.84%
3. (L2 & L3) a. BCl_3 b. $CuCl_2$
   c. $Al(NO_3)_3$ d. Na_2S
   e. CH_4 f. C_4H_{10}

4. (L2 & L3) a. potassium acetate
   b. titanium(IV) chloride

c. calcium iodide

d. ammonium perchlorate

5. **(L1)** a. $(NH_4)_2SO_3$: two NH_4^+ and one SO_3^{2-}

b. $CaCl_2$: one Ca^{2+} and two Cl^-

c. K_2SO_4 : two K^+ and one SO_4^{2-}

d. $Mg_3(PO_4)_2$: three Mg^{2+} and two PO_4^{3-}

6. **(L5)** a. 98.91 g/mol   b. 0.0202 mole

c. 1.22×10^{22} molecules   d. 2.44×10^{22} atoms

7. **(L6)**   % phosphorus

     a. $Ca(H_2PO_4)_2$ 26.47%

     b. $NH_4H_2PO_4$ 26.92%

<u>The second compound is slightly higher.</u>

8. **(L4)** 31 g/mol

9. **(L6)** a. C_5Cl_6 b. $C_{10}Cl_{12}$

10. **(L7)** percent decomposed = 25.5%

CHAPTER 4
PRINCIPLES OF CHEMICAL
REACTIVITY: CHEMICAL REACTIONS

CHAPTER OVERVIEW

Although chemical reactions are a major concern for chemistry courses, beginning students sometimes loose hope that they can ever master so many different equations. Chemists cope with the many different chemical reactions by identifying types of reactions which have similar products or other characteristics. For example, the complete combustion of a hydrocarbon always produces carbon dioxide and water. Thus, if you can recognize that a hydrocarbon is being burned, it isn't hard to predict what the products will be. Once a type has been recognized, it is much easier to predict what products will result from a given set of reactants.

This chapter deals with several reaction types, including four of the most important types of inorganic reactions: precipitation, acid-base, gas-forming, and oxidation-reduction. Once you learn to identify each of these types and know the expected products, it should be much easier to write and use chemical equations.

In order to identify the reaction type, you must first learn to recognize some important classifications of reactants, including acids and bases, gas-forming compounds, and compounds that are oxidized or reduced. Other skills that are important are identifying the ionization products of each soluble reactant, eliminating the spectator ions, and writing the net ionic equation. If you master each of these steps, as presented in this chapter, you should soon be able to write chemical equations for many of the most common chemical reactions.

LEARNING GOALS

1. Know how to balance simple chemical equations. (Sec. 4.2)

2. Recognize nonelectrolytes, strong electrolytes, and weak electrolytes. (Sec. 4.3)

3. Use a table, such as that provided in the text, to predict the solubility of ionic compounds. (Sec. 4.3)

4. Learn the names and formulas of the common acids and bases, as shown in Table 4.1, and be able to predict the products of acid-base reactions. (Sec. 4.4)

5. Write net ionic equations for reactions in aqueous solution. (Sec. 4.5)

6. Recognize and predict the products of acid-base, precipitation, oxidation-reduction, and gas-forming reactions. (Sec. 4.6. 4.7, 4.8, 4.9 and 4.10)

7. Be able to determine the oxidation number for elements in compounds and ions, and use this information to recognize oxidation-reduction reactions. (Sec. 4.10)

ARMCHAIR EXERCISE
(Helpful hints are at the end of the chapter.)

1. Milk is often recommended as a good source of the calcium ion, but sometimes chocolate flavoring be added to the milk to make it more attractive to children. Since chocolate contains oxalic acid, does this seem like a good idea?

2. Suppose that you have three white solids, sugar, salt, and sand. Can you think of a way to identify each of these substances, <u>without tasting them.</u> You may refer to a known sample of the solid for comparison.

3. Geologists commonly identify limestone rocks by dropping acid on the rock surface. If limestone is calcium carbonate, what do you expect to happen when this test is performed? Identify any expected product.

4. Acetic acid and ammonia are each weak electrolytes, but if aqueous solutions of these two chemicals are mixed, the result is a strong electrolyte. Can you explain this? Hint: Look at the product of this reaction.

CONCEPT TEST
(See answers at the end of the chapter.)

1. The substances combined in a chemical reaction are called the _____; the substances produced in a chemical equation are called the _____.

2. A(n) _____ is a substance that increases the concentration of the hydroxide ion, OH^-, when dissolved in pure water.

3. Substances that dissolve in water but do not ionize are called _____.

4. The products of the <u>complete</u> combustion of hydrocarbons are always the compounds _____ and _____.

5. Chemical reactions in which compounds break down into simpler compounds are called _____. Often this occurs due to heating.

6. When a metal carbonate reacts with acid, the products are a(n) _____ and _____.

7. Combustion reactions involve the combination of some reactant (or reactants) with _____ (name the element).

8. Ionic compounds that dissociate completely into ions in water and are good conductors of electricity are called _____ electrolytes.

9. Ions that appear in <u>exactly</u> the same form on the reactant and product sides of an ionic equation are called _____.

10. _____ is the name we give to the class of compounds that produces hydrogen ions when dissolved in water.

11. Write the general net ionic equation for the neutralization reaction of any strong acid with any strong base. _____

12. The endpoint of an acid-base titration is often detected by adding a dye called a(n) _____ that changes color at a hydronium ion concentration as close as possible to the equivalence point.

13. Oxidation-reduction reactions are those involving the transfer of _____.

14. When a substance accepts electrons during a chemical reaction, it is said to be _____. The substance oxidized in a chemical reaction is called the _____ agent.

15. _____ is defined as the charge an atom has, or appears to have, when the electrons of the compound are counted according to a certain set of rules.

STUDY HINTS

1. Recognize that you can change only the coefficients, not the subscripts, when attempting to balance a chemical equation. Changing the subscripts actually changes the compound, whereas changing the coefficients only affects the relative quantities of the substances participating in the process.

2. Be sure that you don't misunderstand the law of conservation of mass by thinking that the number of moles of products must be the same as the number of moles of reactants in order for an equation to be balanced. This is not true! Instead, it is the number of moles of each reactant element that must be balanced by the same number of moles of that element as a product.

3. Students will often confuse the weak base ammonia, NH_3, with the ammonium ion, NH_4^+. Another

pair that is sometimes confused is nitric acid, HNO_3, and ammonia.

4. When trying to determine whether or not a given reaction occurs, it helps to be systematic. First look for reactions such as acid-base or gas formation, since these don't place special solubility requirements on the reactants or products. Only when you have eliminated these possibilities should you begin to worry about the solubilities.

5. Remember that the oxidized substance is the reducing agent, and the reduced substance is the oxidizing agent. Also remember that the oxidizing agent and reducing agent in a chemical equation must both be reactants.

6. When trying to identify the substances oxidized and reduced in a chemical equation, always check for free elements, since they are oxidized or reduced when they participate in a chemical reaction.

PRACTICE PROBLEMS

1. **(L1)** Balance each equation.

a. ____ $Sn(s)$ + ____ $Cl_2(g)$ → ____ $SnCl_4(s)$

b. ____ $NH_3(g)$ + ____ $O_2(g)$ → ____ $NO(g)$ + ____ $H_2O(g)$

c. ____ $C_2H_6(g)$ + ____ $O_2(g)$ → ____ $H_2O(g)$ + ____ $CO_2(g)$

d. ____ $Ca_3(PO_4)_2(s)$ + ____ $C(s)$
    → __ $CaO(s)$ + __ $CO(g)$ + __ $P(s)$

2. **(L2)** Complete the following equations by writing the formula of the product (or products) and balancing. (Assume each reaction goes to completion.)

a. ____ $K(s)$ + ____ $Cl_2(g)$ → _____

b. ____ $C_2H_4(g)$ + ____ $O_2(g)$ → _____ + _____

c. ____ $LiCO_3(s)$ + heat → _____ + _____

3. **(L2 and L4)** Name the following common acids and bases, and indicate whether each is a strong or weak electrolyte.

a. H_2SO_4 _____

b. NH_3 _____

c. CH_3COOH _____

d. H_2CO_3 _____

4. **(L5)** Rewrite each equation, eliminating spectator ions to create a balanced, net ionic equation.

a. $Li(OH)(aq) + HNO_3(aq) \rightarrow LiNO_3(aq) + H_2O(\ell)$

b. $Pb(NO_3)_2(aq) + HCl(aq) \rightarrow PbCl_2(s) + HNO_3(aq)$

5. **(L3)** For each of the following ionic compounds, write the ions formed if it is water soluble or just write poorly soluble for substances that dissolve to only a slight extent.

a. K_2CO_3 _____

b. $CuCl_2$ _____

c. PbS _____

d. $BaSO_4$ _____

6. **(L6)** Complete and balance the following equations. Assume complete neutralization.

a. $KOH(aq) + HCl(aq) \rightarrow$ _____ + _____

b. $NaOH(aq) + H_2SO_4(aq) \rightarrow$ _____ + _____

c. $NH_3(aq) + HNO_3(aq) \rightarrow$ _____ + _____

7. **(L7)** For each ion or compound determine the oxidation number for each element. (Don't forget the sign!)

a. $TiCl_4$ _____

b. SO_4^{2-} _____

c. $HClO_4$ _____

d. SiH_4 _____

e. HPO_4^{2-} _____

f. $HClO_3$ _____

8. **(L6)** Classify each of these reactions as an acid-base reaction, a precipitation reaction, or a gas-forming reaction, then write a balanced, net ionic equation, indicating the states (gas, liquid, or solid) for each product.

a. $NH_3(aq) + HCl(aq) \rightarrow NH_4Cl$

b. $BaCO_3(s) + HCl(aq) \rightarrow H_2O + CO_2 + BaCl_2$

c. $NaF(aq) + BaNO_3(aq) \rightarrow BaF_2 + NaNO_3$

d. $Na_2S(aq) + BiCl_3(aq) \rightarrow Bi_2S_3 + 3\ NaCl$

e. $HClO_4(aq) + LiOH(aq) \rightarrow LiClO_4 + H_2O$

9. **(L7)** Supply the information requested for each reaction.

a. $KNO_2(aq) + H_2SO_4(aq) + KI(aq)$
$$\rightarrow NO(g) + H_2O(\ell) + K_2SO_4(aq) + 1/2\ I_2(aq)$$

element oxidized _____   element reduced _____
oxidizing agent _____   reducing agent _____

b. $Sn(s) + HNO_3(aq) \rightarrow Sn(NO_3)_2(aq) + NO(g) + H_2O(\ell)$

element oxidized _____   element reduced _____
oxidizing agent _____   reducing agent _____

PRACTICE PROBLEM SOLUTIONS

1. a. $Sn(s) + 2\ Cl_2(g) \rightarrow SnCl_4(s)$

   b. $4\ NH_3(g) + 5\ O_2(g) \rightarrow 4\ NO(g) + 6\ H_2O(g)$

   c. $2\ C_2H_6(g) + 7\ O_2(g) \rightarrow 6\ H_2O(g) + 4\ CO_2(g)$

   d. $Ca_3(PO_4)_2(s) + 5\ C(s)$
   $$\rightarrow 3\ CaO(s) + 5\ CO(g) + 2\ P(s)$$

2. a. $2\ K(s) + Cl_2(g) \rightarrow 2\ KCl(s)$

   b. $C_2H_4(g) + 3\ O_2(g) \rightarrow 2\ CO_2(g) + 2\ H_2O(g)$

   c. $Li_2CO_3(s) + heat \rightarrow Li_2O(s) + CO_2(g)$

46

3. a. sulfuric acid $\qquad$ strong electrolyte
   b. ammonia $\qquad$ weak electrolyte
   c. acetic acid $\qquad$ weak electrolyte
   d. carbonic acid $\qquad$ weak electrolyte

4. a. $OH^-(aq) + H^+(aq) \rightarrow H_2O(\ell)$

   b. $Pb^{2+}(aq) + 2\ Cl^-(aq) \rightarrow PbCl_2(s)$

5. a. $K_2CO_3 \rightarrow 2\ K^+ + CO_3^{2-}$

   b. $CuCl_2 \rightarrow Cu^{2+} + 2\ Cl^-$

   c. PbS and d. $BaSO_4$ are insoluble

6. a. $KOH(aq) + HCl(aq) \rightarrow KCl(aq) + H_2O(\ell)$

   b. $2\ NaOH(aq) + H_2SO_4(aq) \rightarrow Na_2SO_4(aq) + 2\ H_2O(\ell)$

   c. $2\ NH_3(aq) + HNO_3(aq) \rightarrow NH_4NO_3(aq) + H_2O(\ell)$

7. a. Ti +4, Cl −1   b. S +6, O −2
   c. H +1, Cl +7, O −2   d. Si +4, H −1
   e. H +1, P +5, O −2   f. H +1, Cl +5, O −2

8. a. acid-base

$NH_3(aq) + H^+(aq) \rightarrow NH_4^+(aq)$

b. gas forming

$BaCO_3(s) + 2\ H^+(aq) \rightarrow H_2O(\ell) + CO_2(g) + Ba^{2+}(aq)$

c. precipitation

$2\ F^-(aq) + Ba^{2+}(aq) \rightarrow BaF_2(s)$

d. precipitation

$3\ S^{2-} + 2\ Bi^{3+}(aq) \rightarrow Bi_2S_3(s)$

e. acid-base
$H^+(aq) + OH^-(aq) \rightarrow H_2O(\ell)$

9. For equation a
   element oxidized: iodine (−1 to 0)
   element reduced: nitrogen (+3 to +2)
   oxidizing agent: KNO_2 reducing agent: KI

9. (cont.) For equation b
element oxidized: tin (0 to +2)
element reduced: nitrogen (+5 to +2)
oxidizing agent: HNO_3 reducing agent: Sn

PRACTICE TEST (20 Minutes)
(Answers are provided at the end of the chapter.)
1. Balance the following equations that describe chemical processes used in industry.

a. formation of dilute phosphoric acid

____ $P_4O_{10}(s)$ + ____ $H_2O(\ell)$ → ____ $H_3PO_4(aq)$

b. catalyzed reduction of nitrogen oxide emissions

__ $NO(g)$+ __ $NO_2(g)$+ __ $NH_3(g)$
→ ____ $N_2(g)$+ ____ $H_2O(g)$

2. Complete and balance the equations for each of the following reactions.

a. $BaCO_3(s)$ + heat → _____ + _____

b. $Li(s)$ + $F_2(g)$ → _____

c. $C_4H_{10}(g)$ + $O_2(g)$ → _____ + _____

d. $K_2CO_3(s)$ + $HCl(aq)$ → _____ + _____

3. For each of the following ionic compounds, write the ions formed if it is water soluble or just write poorly soluble for substances that dissolve to only a slight extent.

a. CaF_2 _____   b. $AgNO_3$ _____

c. $MgCrO_4$ _____   d. $CuSO_4$ _____

4. For each ion or compound determine the oxidation number (with sign) for each element.

a. H_3AsO_3 _____   b. PO_4^{3-} _____

48

5. Rewrite this equation, eliminating spectator ions, and balance the resulting net ionic equation.

$$K_2S \quad + \quad CdCl_2 \quad \rightarrow \quad CdS \quad + \quad KCl$$

6. Name the following acids and bases, and indicate whether each is a strong or weak electrolyte.

a. $HClO_4$ _____

b. NaOH _____

7. Classify each of these reactions as an acid-base reaction, a precipitation reaction, or a gas-forming reaction. Show the states (gas, liquid, or solid) for each product, then write and balance the net ionic equation.

a. $LiOH(aq) + HNO_3(aq) \rightarrow LiNO_3 + H_2O$

b. $MgCO_3(s) + HNO_3(aq) \rightarrow H_2O + CO_2 + Mg(NO_3)_2$

c. $Sr(NO_3)_2(aq) + Na_2SO_4(aq) \rightarrow NaNO_3 + SrSO_4$

8. Supply the information requested for each reaction.

a. $H_2O(\ell) \quad + \quad Cl_2(aq) \quad \rightarrow \quad HCl(aq) \quad + \quad HOCl(aq)$

element oxidized _____ element reduced _____

oxidizing agent _____ reducing agent _____

b. $N_2H_4(aq) + O_2(g) \rightarrow N_2(g) + H_2O(\ell)$

element oxidized _____ element reduced _____

oxidizing agent _____ reducing agent _____

COMMENTS ON ARMCHAIR EXERCISES

1. Have you checked the solubility of calcium oxalate? Does it seem likely that calcium will be available when oxalate is also present?

2. The sand is easy, since it is insoluble in

water, whereas the other two are soluble. There are several ways to distinguish between sugar and salt. If you put one of the solids on a spoon and heat gently over a candle, the sugar will char, but the salt will not. If you have a way to test conductivity, you might dissolve both in water and determine which solution conducts an electric current. A strong magnifying glass (30x) will allow you to see that the crystals of the two solids have different shapes. Can you think of other ways to identify these two solids?

3. This is a gas-forming reaction. The equation is

$$CaCO_3(s) + 2 H^+ \rightarrow H_2O(\ell) + CO_2(g) + Ca^{2+}$$

4. Acetic acid and ammonia are both weak electrolytes, but when they combine, they produce a salt, ammonium acetate, which is a strong electrolyte.

CONCEPT TEST ANSWERS

1. reactants, products   2. base
3. nonelectrolytes   4. CO_2, H_2O
5. decomposition
6. metal oxide, carbon dioxide
7. oxygen   8. strong
9. spectator ions   10. acid
11. OH^- + H^+ $\rightarrow$ H_2O
12. indicator   13. electrons
14. reduced, reducing agent
15. oxidation number

PRACTICE TEST ANSWERS

1. (**L1**) a. $P_4O_{10}(s) + 6 H_2O(\ell) \rightarrow 4 H_3PO_4(aq)$

   b. $NO(g) + NO_2(g) + 2 NH_3(g) \rightarrow 2 N_2(g) + 3 H_2O(g)$

2. (**L2**) a. $BaCO_3(s) + heat \rightarrow BaO(s) + CO_2(g)$

   b. $2 Li(s) + F_2(g) \rightarrow 2 LiF(s)$

   c. $2 C_4H_{10}(g) + 13 O_2(g) \rightarrow 8 CO_2(g) + 10 H_2O(\ell)$

d. $K_2CO_3(s) + 2\ HCl(aq)$
$$\rightarrow 2\ KCl(aq) + H_2O(\ell) + CO_2(g)$$

3. **(L3)** a. $AgNO_3(aq) \rightarrow Ag^+(aq) + NO_3^-(aq)$

   b. $CuSO_4(aq) \rightarrow Cu^{2+}(aq) + SO_4^{2-}(aq)$

   c. CaF_2 and c. $MgCrO_4$ are insoluble

4. **(L7)** a. H +1, As +3, and O −2   b. P +5 and O −2

5. **(L5)** $S^{2-}(aq) + Cd^{2+}(aq) \rightarrow CdS(s)$

6. **(L4)** a. perchloric acid   strong electrolyte
   b. sodium hydroxide   strong electrolyte

7. **(L6)** a. acid-base

$OH^-(aq) + H^+(aq) \rightarrow H_2O(\ell)$

b. gas forming

$MgCO_3(s) + 2\ H^+(aq) \rightarrow H_2O(\ell) + CO_2(g) + Mg^{2+}(aq)$

c. precipitation

$Sr^{2+}(aq) + SO_4^{2-}(aq) \rightarrow SrSO_4(s)$

8. **(L7)** For equation a
      element oxidized: chlorine (0 to +1)
      element reduced: chlorine (0 to −1)
      oxidizing agent: Cl_2 reducing agent: Cl_2
This is a special kind of oxidation-reduction reaction. The same element is both oxidized and reduced, and so it is called a disproportionation reaction.

b. For equation b
      element oxidized: nitrogen (+2 to 0)
      element reduced: oxygen (0 to −2)
      oxidizing agent: O_2 reducing agent: N_2H_4

CHAPTER 5
STOICHIOMETRY

CHAPTER OVERVIEW

In the previous chapter you learned to write balanced chemical equations for chemical reactions. Balanced equations are important because they provide quantitative relationships between the amounts of products and reactants involved in a chemical reaction, relationships that may be stated in terms of atoms and molecules, moles, or grams. When the reactants are not in the correct mole ratio it's first necessary to determine the limiting reactant, that is, the reactant which controls the maximum amount of reaction that can occur.

Stoichiometric calculations are used for many different types of chemical calculations, including the analysis of mixtures and the determination of the formulas of compounds. Another kind of calculation which is especially useful is the determination of percent yield. The percent yield compares the actual yield with the theoretical yield, that is, the amount of compound prepared in a given experiment with the maximum amount that could be obtained from the available amounts of starting reactants. The higher the percent yield, the more efficiently the reaction was performed.

When using mole relations for reactions that occur in solution, it is usually more convenient to use the concentration of the compounds involved as a basis for the calculations. An important example of this type of stoichiometry is the determination of the amount of a particular material in an unknown solution, a process called titration.

LEARNING GOALS

1. Use a balanced chemical equation to write the stoichiometric coefficients for a chemical reaction and then calculate the relationships between the moles or mass of products and reactants. (Sec. 5.1)

2. In some stoichiometric calculations the amount of one of the reactants causes that reagent to be the controlling factor in determining the extent to which the reaction will occur. This is called a limiting reagent problem. It's necessary to be able to recognize this type of problem, to identify the limiting reagent, and to use the amount of limiting reagent to perform the calculations described in learning goal 1. (Sec. 5.2)

3. Understand what is meant by actual yield, theoretical yield, and percent yield, and use the equation

$$\text{percent yield} = \frac{\text{actual yield x 100}}{\text{theoretical yield}}$$

to do simple calculations. (Sec. 5.3)

4. Some applications of the stoichiometric principles discussed in the chapter are the use of combustion analysis to determine the composition of a mixture or the empirical and molecular formulas. Know how to work these problems. (Sec. 5.4)

5. Solution concentrations are frequently expressed in terms of molarity, where

$$\text{molarity (M)} = \frac{\text{moles of solute}}{\text{volume of solution (L)}}$$

Use this equation to do simple solution calculations, including those required to prepare a solution of a given concentration either by dissolving the solute directly in a solvent or by dilution of a previously known solution. (Sec. 5.5)

6. Be able to use a balanced equation and the molarity of the reactants to perform solution stoichiometry calculations for titration and precipitation reactions. (Sec. 5.6)

ARMCHAIR EXERCISES

1. Fenster, Harpp, and Schwarcz (*J.Chem.Ed.* <u>1987</u>, *64*, 894) describe an experiment involving the combustion of gaseous mixtures of oxygen and a simple hydrocarbon called propane. DO NOT ATTEMPT TO DO THIS EXPERIMENT YOURSELF! If a balloon

containing only propane is ignited by means of a candle on the end of a long stick, there is relatively little noise. If oxygen is mixed with the propane before it is ignited, the noise is much louder. As balloons are detonated with increasing amounts of oxygen, but the same amount of propane, the noise becomes louder until a concentration is used that produces the biggest bang. Adding further oxygen beyond that point does not make the noise any greater. Can you explain this observation in terms of the ideas in this chapter?

2. One way to review many of the concepts discussed thus far is to work what is called an open-ended problem. Suppose that you have a 27.0 gram sample of water. Do as many calculations as possible based on this quantity of water, using the information on a standard periodic table in addition to any standard relationships that you have learned. For example, if you remember that the density of water is 1.00 gram/milliliter, you should be able to calculate the volume of this water sample. How many different calculations can you do here?

CONCEPT TEST
(Answers are provided at the end of the chapter.)
1. When one reactant in a chemical reaction is present in an amount less than that required by the stoichiometry of the equation, that reactant is called the _____ and it will determine the quantity of products formed.

2. The maximum amount of product that can be isolated from a given quantity of reactants is called the _____.

3. The efficiency of a chemical reaction is often evaluated in terms of the _____ , that is, the actual yield times 100 divided by the theoretical yield.

4. If the initial moles of methane (natural gas)

reacting is known, to calculate how many moles of oxygen are required to react with it in the balanced equation

$$CH_4 + 2 O_2 \rightarrow CO_2 + 2 H_2O$$

the stoichiometric factor is _____.

5. _____ is name given to the procedure used to determine the exact concentration of an acid, base, or other reagent, usually in preparation for using that substance in a titration.

6. When a compound containing only carbon and hydrogen is completely burned in air, the products are _____ and _____.

7. The mole ratio of each element in a compound is called the _____.

8. _____ is the name of a standard concentration unit, which is moles of solute per liter of solution.

9. When the formula of a compound is written in square brackets, this means _____.

10. The dilution method for preparing a solution of a specific concentration consists of adding _____ to a more concentrated solution having a known concentration.

11. Dilution problems are based on the idea that the number of _____ of solute in the initial solution must equal the number of _____ of solute in the final solution.

12. _____ is the industrial chemical produced in the greatest quantity per year in this country.

13. _____ is the determination of the identity of the components in a mixture.

14. The determination of the quantity of a given constituent in a mixture is called _____.

15. That point in an acid-base titration at which the number of moles of OH^- added is equal to the number of moles of H_3O^+ present is called the

_____.

16. A(n) _____ is a dye which is added to titrations to indicate when the reaction is complete.

STUDY HINTS

1. When you begin to work on a stoichiometry problem your first step should always be to make sure that any chemical equations are complete and balanced. Especially in this type of problem, an equation that hasn't been balanced is a mistake waiting to happen.

2. It is very important to realize that the mole ratio in the balanced equation is the key to doing most of the problems in this chapter. As the textbook points out, most of these problems can be thought of in three steps: first, convert whatever you are given into moles; second, use the stoichiometric ratio to convert to moles of another substance in the balanced equation; and third, convert from moles of that new substance to whatever units are requested.

3. Notice that the limiting reagent is not necessarily the reactant present in the smallest amount; you must also consider the stoichiometric factor. The reagent present in the largest amount may be the limiting reagent if there is not enough of it to satisfy the requirements of the balanced equation. Watch for a practice problem that illustrates this.

4. When you learn the definition of molarity, be sure to learn moles <u>of solute</u> per liter <u>of</u>

solution, not just moles per liter. Remember that when preparing solution, liters of solvent added is not always the same as the final volume of the solution. In addition, you will soon learn another concentration unit with a name and definition similar to molarity. If you learn the complete definition for molarity now, you are less likely to be confused at that time.

5. Dilution problems should be fairly easy once you memorize the formula, but there is one frequent point of confusion. You must read carefully to distinguish between the volume of water added and the volume of the final solution. The three sections of practice problem 6 show how this type of question can be asked several different ways.

6. One of the more confusing problem types you will encounter in this chapter is the determination of weight percent by means of a titration. In these problems, the mass of the sample is usually given, and enough information is provided to calculate the mass of some component of that sample. Unless you read the problem carefully it can be easy to confuse the mass of the sample with the mass of the component being measured. Remember that if the sample contained nothing but that component, there would be no need to do an analysis.

PRACTICE PROBLEMS

1. (L1) Answer the following question based on the balanced equation
$$Al_4C_3(s) + 12\ H_2O(\ell) \rightarrow 4\ Al(OH)_3(aq) + 3\ CH_4(g)$$

a. How many moles of water are necessary to react completely with 6.42 moles of Al_4C_3?

b. How many moles of $Al(OH)_3$ will result from the complete reaction of 265 grams of water with excess aluminum carbide, Al_4C_3?

c. How many grams of H_2O are required to prepare 128 grams of CH_4 using the above reaction?

2. (L2) Crude Titanium metal is prepared commercially according to the balanced equation

$$TiCl_4 \ + \ 2 \ Mg \ \rightarrow \ 2 \ MgCl_2 \ + \ Ti$$

If 40.0 kilograms of magnesium is reacted with 85.2 kilograms of titanium chloride, (a) what is the limiting reagent, and (b) how many grams of titanium metal will be formed?

3. (L3) In the previous problem, the theoretical yield of titanium metal is 21.5 kilograms. If the actual yield of titanium metal in the experiment was 10.4 kilograms, what was the percent yield for this reaction?

4. (L4) A certain compound that consists only of carbon and hydrogen is burned in air, producing 8.80 grams of carbon dioxide and 7.20 grams of water. What is the empirical formula of this compound?

5. (L5) a. What is the molarity of the solution that results when 20.0 mL of water is added to 10.0 mL of 0.200 molar NaOH?
b. What is the molarity of the solution that results when 10.0 mL of a 0.200 molar solution of NaOH is diluted with water to a final volume of 20.0 mL?
c. How much water must be added to 10.0 mL of a 0.200 molar solution of NaOH to produce a final solution that is 0.080 molar in NaOH?

6. (L5) How many liters of 0.444 molar KOH solution can be prepared from 127 grams of KOH?

7. (L6) How many mL of 0.20 molar barium chloride will be required to react completely with 350 mL of 0.15 molar sulfuric acid according to the equation (balance first if necessary)

$$BaCl_2 \; + \; H_2SO_4 \; \rightarrow \; BaSO_4 \; + \; HCl$$

8. (L6) If 27.9 mL of KOH solution is required to titrate 25.0 mL of 0.150 molar HCl to the endpoint, what is the molarity of the KOH solution?

9. (L6) Calculate the percent by weight of acetic acid in a vinegar sample if it requires 35.18 mL of 0.420 M NaOH to neutralize the acid in 25.0 mL of a sample of vinegar. You may assume all of the acidity is due to acetic acid, $HC_2H_3O_2$, and that the density of the vinegar is 1.00 g/mL.

10. **(L8)** a 2.00 gram sample of tin(II) chloride is titrated in acidic solution with 0.100 molar potassium permanganate. How many mL of potassium permanganate will be required if the balanced equation for the reaction is

$$16 \ H^+ + 5 \ Sn^{2+} + 2 \ MnO_4^- \rightarrow 2 \ Mn^{2+} + 5 \ Sn^{4+} + 8 \ H_2O$$

PRACTICE PROBLEM SOLUTIONS

1. a. According to the balanced equation, 12 moles of water is necessary to react completely with 1 mole of Al_4C_3.

$$\text{moles of water} = 6.42 \ \text{mol} \ Al_4C_3 \ \times \ \frac{12 \ \text{mol water}}{1 \ \text{mol} \ Al_4C_3}$$

moles water = 77.0 moles

b. First, determine the number of moles of water.

$$\text{moles of water} = \frac{265 \ g}{18.02 \ g/mol}$$

moles of water = 14.71 mol

From the equation 12 mol of water produces 4 mol of $Al(OH)_3$, so using this stoichiometric factor

$$\text{mol of } Al(OH)_3 = 14.71 \ \text{mol water} \ \times \ \frac{4 \ \text{mol aluminum hydroxide}}{12 \ \text{mol water}}$$

moles of Al_4C_3 = <u>4.90 moles</u>

c. As before, the first step is to determine the moles of CH_4

$$\text{moles of } CH_4 = \frac{128 \text{ g}}{16.04 \text{ g/mol}} = 7.975 \text{ moles}$$

From the balanced equation, 12 moles of water are required to produce 3 moles of methane.

$$\text{moles of } H_2O = 7.975 \text{ moles } CH_4 \text{ x } \frac{12 \text{ mol water}}{3 \text{ mol } CH_4}$$

moles of H_2O = 31.90 mol

Finally, determine the grams of water

grams of water = 31.90 moles x 18.02 g/mol

<u>grams of water = 575 g</u>

2. a. The first step in determining the limiting reagent is to calculate the number of moles of each reagent available.

$$\text{moles } TiCl_4 = \frac{85.2 \text{ kg x } 1000 \text{ g/kg}}{189.7 \text{ g/mol}} = 449.2 \text{ moles}$$

$$\text{moles Mg} = \frac{40.0 \text{ kg x } 1000 \text{ g/kg}}{24.31 \text{ g/mol}} = 1645 \text{ moles}$$

Using the balanced equation, calculate the ratio of moles required.

$$\text{ratio of moles required} = \frac{2 \text{ moles Mg}}{1 \text{ mol } TiCl_4} = 2.0$$

Compare this with the moles available

$$\text{ratio of moles available} = \frac{1645 \text{ moles Mg}}{449.2 \text{ moles } TiCl_4} = 3.66$$

The "ratio of moles required" is smaller than the "ratio of moles available" so the reagent in the denominator is the limiting reagent.
    Titanium(IV) Chloride is the limiting reagent.

b. The balanced equation indicates that one mole of $TiCl_4$ will form one mole of Ti, so the next step is easy.

moles Ti formed = moles of $TiCl_4$ = 449.2 moles

Now, convert the moles of Ti to grams of Ti.

mass of Ti = 449.2 moles of Ti x 47.88 g/mol

mass of Ti = 2.15×10^4 g

3. Remember that the definition of percent yield is

$$\text{percent yield} = \frac{\text{actual yield x 100}}{\text{theoretical yield}}$$

Substituting into this equation

$$\text{percent yield} = \frac{10.4 \text{ kg x 100}}{21.5 \text{ kg}}$$

percent yield = 48.4 %

4. All of the carbon in the original compound has been converted into carbon dioxide, and all of the hydrogen into water. Therefore, it's apparent that

moles of carbon = moles of carbon dioxide formed

moles of hydrogen = 2 x moles of water formed

Next calculate the moles of water and carbon dioxide.

$$\text{mol } CO_2 = \frac{8.80 \text{ g}}{44.01 \text{ g/mol}} = 0.200 \text{ moles}$$

$$\text{mol } H_2O = \frac{7.20 \text{ g}}{18.02 \text{ g/mol}} = 0.400 \text{ moles}$$

Determine the moles of hydrogen and carbon

$$mol\ C = \frac{1\ mol\ C}{1\ mol\ CO_2} \times 0.2000\ mol\ CO_2$$

$$= 0.2000\ mol\ of\ C$$

$$mol\ H = \frac{1\ mol\ H}{2\ mol\ H_2O} \times 0.4000\ mol\ H_2O$$

$$= 0.8000\ mol\ of\ H$$

Divide the number of moles of each element by the smallest number of moles, that is, 0.200

$$\frac{0.8000\ mol\ H}{0.2000\ mol\ C} = \frac{4\ mol\ H}{1\ mol\ C}$$

The empirical formula is CH_4

5. Before you do any of the parts of this problem, read all of them carefully. Notice that even though the wording is very similar, there are significant differences among the three problems. In the first case, you are told how much water is added; the second problem is based on a final volume of solution, and the last part asks how much water should be added. The differences are subtle but important. Be sure that you can recognize how each type of problem is worked.

a. This is a case where a data table may be helpful.

	initial	final
volume	10.0 mL	10.0 + 20.0 = 30.0 mL
concentration	0.200 M	X

Next substitute into the dilution equation.

$$M_c V_c = M_d V_d$$

$$0.200\ M \times 10.0\ mL = X \times 30.0\ mL$$

$$X = 0.0667\ M$$

b. Again, we will start with a data table.

	initial	final
volume	10.0 mL	20.0 mL
concentration	0.200 M	X

Notice how the changed wording has affected the problem. Next substitute into the equation.

$$M_c V_c = M_d V_d$$

0.200 M x 10.0 mL = X x 20.0 mL

$$\underline{X = 0.100\ M}$$

c. As before, changing the statement of the problem does affect the data we will use.

	initial	final
volume	10.0 mL	X
concentration	0.200 M	0.080 M

Next substitute into the dilution equation.

$$M_c V_c = M_d V_d$$

0.200 M x 10.0 mL = 0.080 M x X

$$X = 25.0\ mL$$

But this result is the <u>total</u> final volume, <u>not</u> the amount of water added. To find the amount of water added, subtract the initial volume, 10.0 mL.

Water added = 25.0 mL - 10.0 mL

<u>Water added = 15.0 mL</u>

6. Start with the defining equation for molarity

$$\text{molarity} = \frac{\underline{\text{moles of solute}}}{\text{volume of solution}}$$

Find the moles of KOH

$$\text{moles KOH} = \frac{\text{grams KOH}}{\text{molar mass}} = \frac{127 \text{ g}}{56.11 \text{ g/mol}} = 2.263 \text{ mol}$$

Rearrange the molarity equation and substitute the known values.

$$\text{liters} = \frac{2.26 \text{ mol}}{0.444 \text{ M}} = \underline{5.10 \text{ Liters}}$$

7. First, write the balanced equation.

$$BaCl_2 \quad + \quad H_2SO_4 \quad \rightarrow \quad BaSO_4 \quad + \quad 2 \text{ HCl}$$

Next, calculate the moles of H_2SO_4 to be used.

$$0.350 \text{ L } H_2SO_4 \times \frac{0.15 \text{ mol}}{1.00 \text{ L}} = 0.0525 \text{ mol } H_2SO_4$$

Now use the balanced equation to find the mole relationship between barium chloride and sulfuric acid.

$$\text{mol } BaCl_2 = 0.0525 \text{ mol } H_2SO_4 \times \frac{1 \text{ mol } BaCl_2 \text{ required}}{1 \text{ mol } H_2SO_4 \text{ available}}$$
$$= 0.0525 \text{ mol } BaCl_2$$

Finally, calculate the mL of $BaCl_2$ needed.

mol $BaCl_2$ needed =

$$0.0525 \text{ mol } BaCl_2 \text{ required} \times \frac{1.00 \text{ L solution}}{0.20 \text{ mol } BaCl_2}$$

$$= 0.263 \text{ L } BaCl_2$$

or

$$= \underline{260 \text{ mL of } BaCl_2}$$

8. First, write the balanced equation.

$$KOH \quad + \quad HCl \quad \rightarrow \quad KCl \quad + \quad H_2O$$

Determine the moles of acid.

$$0.0250 \text{ L HCl} \times \frac{0.150 \text{ mole HCl}}{1.00 \text{ L HCl}} = 0.00375 \text{ mol HCl}$$

Calculate the molarity of base required for 27.9 mL of base to neutralize 0.00375 moles of acid.

$$0.0279 \text{ L NaOH} \times \frac{X \text{ mole NaOH}}{1.00 \text{ L NaOH}} = 0.00375 \text{ mol NaOH}$$

molarity of NaOH = 0.134 M

9. Again, start with a balanced equation

$$HC_2H_3O_2 \quad + \quad NaOH \quad \rightarrow \quad NaC_2H_3O_2 \quad + \quad H_2O$$

Next, determine how many moles of base reacted.

$$\text{mol NaOH} = \frac{35.18 \text{ mL}}{1000 \text{ mL/L}} \quad \times \quad \frac{0.420 \text{ mol}}{1.000 \text{ L}}$$

mol NaOH = 0.01478 mol

Next, use relationship from the balanced equation to find how many moles of acid would be necessary to react completely with this amount of NaOH.

$$\text{mol } HC_2H_3O_2 = 0.01478 \text{ mol NaOH} \times \frac{1 \text{ mol } HC_2H_3O_2}{1 \text{ mol NaOH}}$$

mol $HC_2H_3O_2$ = 0.01478 moles

Convert the moles to grams

$$\text{mass of } HC_2H_3O_2 = 0.01478 \text{ mol} \times 60.06 \text{ g/mol}$$

$$= 0.8874 \text{ grams}$$

Since the density of the vinegar is 1.00 g/mL, the mass will be numerically equal to the volume.

mass of vinegar = 25.0 g

The equation for weight percent is

$$\text{weight percent} = \frac{\text{grams of acetic acid} \times 100}{\text{grams of sample (vinegar)}}$$

$$\text{weight percent} = \frac{0.8874 \text{ g} \times 100}{25.0 \text{ grams}}$$

percent of $HC_2H_3O_2$ = 3.55%

10. The balanced equation is already provided.

$$16 \text{ H}^+ + 5 \text{ Sn}^{2+} + 2 \text{ MnO}_4^- \rightarrow 2 \text{ Mn}^{2+} + 5 \text{ Sn}^{4+} + 8 \text{ H}_2\text{O}$$

Start by calculating the moles of $SnCl_2$.

$$\text{moles of } SnCl_2 = \frac{2.00 \text{ g tin(II) chloride}}{189.6 \text{ g/mol}}$$

$$= 0.01055 \text{ mol } SnCl_2$$

Now use the balanced equation to determine how many moles of $KMnO_4$ will be needed.

$$\text{mol } MnO_4^- = 0.01055 \text{ mol } SnCl_2 \times \frac{2 \text{ mol } MnO_4^-}{5 \text{ mol } Sn^{2+}}$$

$$= 0.004219 \text{ mol } MnO_4^-$$

Finally determine the volume of MnO_4^- needed.

$$\text{mol } MnO_4^- = 0.004219 \text{ mol } MnO_4^- \times \frac{1.00 \text{ L solution}}{0.100 \text{ mol } MnO^{4-}}$$

$$= \underline{0.04219 \text{ L } MnO_4^-}$$

That is, 42.2 mL of permanganate is required.

PRACTICE TEST (50 minutes)

1. Pure silicon is produced for transistors by reacting silicon tetrachloride with magnesium according to the equation

$$SiCl_4(g) + 2 \text{ Mg}(s) \rightarrow 2 \text{ MgCl}_2(s) + Si(s)$$

How many moles of magnesium are required to react completely with 34 grams of $SiCl_4$?

2. Nitric acid can be produced by reacting nitrogen dioxide with water according to the equation

$$3 \text{ NO}_2(g) + \text{H}_2\text{O}(\ell) \rightarrow 2 \text{ HNO}_3(aq) + \text{NO}(g)$$

How many grams of HNO_3 can be produced by reacting 13.8 grams of NO_2 with an excess of water?

3. Fibers of boron carbide are used in some bullet-proof protective clothing. Boron carbide can be synthesized from boron oxide by means of the following two step process

$$B_2O_3(s) + 3\ C(s) + 3\ Cl_2(g) \rightarrow 6\ CO(g) + 2\ BCl_3(g)$$

$$4\ BCl_3(g) + 6\ H_2(g) + C(s) \rightarrow B_4C(s) + 12\ HCl(g)$$

How many moles of B_4C could be produced from the complete reaction of 250 moles of B_2O_3 according to the sequence of reactions given above?

4. Iron can be obtained in a blast furnace by the reaction of the ore hematite with carbon in a series of reactions that may be represented by the overall equation

$$2\ Fe_2O_3\quad +\quad 3\ C\quad \rightarrow\quad 3\ CO_2\quad +\quad 4\ Fe$$

If a company has a stockpile of $3.27x10^6$ kg of hematite and $5.85x10^5$ kg of carbon, how many grams of iron can be produced?

5. Xenon and fluorine are reacted to form 37.9 grams of xenon tetrafluoride. Based on the masses of the starting materials, the expected yield of xenon tetrachloride was 64.5 grams. What is the percent yield of this process?

6. A certain compound that consists only of carbon and hydrogen is burned in air, producing 13.2 grams of carbon dioxide and 7.20 grams of water. What is the empirical formula of this compound?

7. Suppose that you wish to prepare 265 mL of 0.500 molar sulfuric acid. How many grams of H_2SO_4 would be required to prepare this solution?

8. How many mL of 0.200 molar HCL would be necessary to exactly neutralize 5.00 grams of $Ba(OH)_2$?

9. a. How much water must be added to $2.50x10^2$ mL of 2.60 M HCl to produce a solution of HCl having a concentration of 0.500 M?
b. Calculate the molar concentration of an HCl

solution formed by adding 120.0 mL of water to 10.0 mL of 5.20 M HCl.

10. If sodium metal is placed in water it will react according to the balanced equation

$$2 \ Na(s) + 2 \ H_2O(\ell) \rightarrow 2 \ NaOH(aq) + H_2(g)$$

Calculate the molar concentration of sodium hydroxide formed in a reaction of this type if it is known that 0.020 grams of hydrogen was generated and the total volume of water after the reaction was 150 mL.

11. A 1.46 gram sample of iron ore is dissolved in acid and reduced so that all of the iron present is in the +2 oxidation state. The resulting solution is titrated to the endpoint with 19.8 mL of 0.31 molar cerium(IV) ion according to the equation

$$2 \ Fe^{2+} \ + \ Ce^{4+} \ \rightarrow \ 2 \ Fe^{3+} \ + \ Ce^{2+}$$

What is the percent of iron in the original sample?

CONCEPT TEST ANSWERS

1. limiting reagent   2. theoretical yield
3. percent yield   4. two
5. standardization   6. water, carbon dioxide
7. empirical formula 8. Molarity
9. concentration of the compound
10. water (or solvent)
11. moles, moles   12. sulfuric acid
13. qualitative analysis
14. quantitative analysis
15. equivalence point 16. indicator

COMMENTS ON ARMCHAIR EXERCISES

1. If you write the balanced equation for this reaction, you will find that the stoichiometric ratio is one of propane for each five moles of oxygen. The equation is

$$C_3H_8(g) \ + \ 5 \ O_2(g) \ \rightarrow \ 3 \ CO_2(g) \ + \ 4 \ H_2O(g)$$

Does this help you to understand why the one to five ratio gave the loudest noise?

2. How many different calculations did you do? Just a few of the possibilities are

    moles of water
    number of water molecules
    moles of hydrogen atoms
    number of hydrogen atoms
    moles of oxygen atoms

or if you're really creative, you might calculate how many grams of hydrogen gas would be required to produce this amount of water, etc.

PRACTICE TEST ANSWERS

1. (L1) 0.40 moles   2. (L1) 12.6 g
3. (L1) 125 moles   4. (L2) 2.29×10^9 grams
5. (L3) 58.8%   6. (L4) C_3H_8
7. (L5) 13.0 grams   8. (L6) 292 mL
9. (L5) a. 1050 mL   b. 0.400 M
10. (L6) 0.13 M   11. (L6) 47%

SPECIAL SECTION I
NORMALITY AND EQUIVALENTS

In some cases when the same type of stoichiometry calculation is done repeatedly, it is convenient to use a concentration unit called **Normality**. The idea behind this concentration unit is that if the amount of each reactant is measured relative to a set amount of some standard reactant, it will be easier to calculate the stoichiometric relationships. This standard amount is called a *chemical equivalent*, because it provides a way to identify quantities of reactants that will have equivalent chemical reactivities.

As will be seen, this approach tends to make the easier problems a little harder, but it also can simplify some rather difficult problems. Since the definition of Normality is not the same for acid-base as for oxidation-reduction reactions, these situations will be discussed separately.

SECTION 1. CHEMICAL EQUIVALENTS AND ACID-BASE TITRATIONS

According to one of the most commonly used acid-base definitions, an acid is a substance that donates protons, that is hydrogen ions, and a base is a substance that accepts protons. This suggests that the amount of acids and bases can be related to the ability to produce or react with one mole of hydrogen (or hydronium) ions. This is the basis of the definition of the chemical equivalent in acid-base reactions. *For acid-base reactions, a* **chemical equivalent** *is defined as the amount of an acid or base that will react with or produce one mole of hydrogen ions.*

For most common acids, the number of equivalents provided by each mole of compound is determined from the ionization equation. For example, compare hydrochloric and sulfuric acids:

$$HCl \rightarrow H^+ + Cl^-$$

$$H_2SO_4 \rightarrow 2\ H^+ + SO_4^{2-}$$

72

One mole of hydrochloric acid will produce one mole of hydrogen ions, or *one equivalent*. Thus, for HCl there is one equivalent per mole. One mole of sulfuric acid will produce two moles of hydrogen ions and so has two equivalents per mole.

It is also easy to determine the number of equivalents per mole for most common bases. Since the neutralization reaction between hydrogen ion and hydroxide ion requires one mole of hydrogen ion for each mole of hydroxide,

$$H^+(aq) \quad + \quad OH^-(aq) \quad \rightarrow \quad H_2O(\ell)$$

one chemical equivalent of a base is the amount that will produce one mole of hydroxide ions.

As before, in order to determine equivalents per mole for bases it is necessary to examine the ionization equations. For example, compare sodium hydroxide and calcium hydroxide.

$$NaOH \quad \rightarrow \quad Na^+ \quad + \quad OH^-$$

$$Ca(OH)_2 \quad \rightarrow \quad Ca^{2+} \quad + \quad 2\ OH^-$$

The number of equivalents per mole for bases is determined by the number of moles of hydroxide ion produced per mole of base. For sodium hydroxide, there is one equivalent per mole, and for calcium hydroxide, there are two equivalents per mole.

It has probably become apparent that one way to find equivalents per mole is to simply count the number of ionizable hydrogens or hydroxides in the formula of the compound. Frequently this works, but there are a number of cases, like ammonia, NH_3, where it's essential to refer to the ionization equation when determining equivalents per mole.

$$NH_3 \quad + \quad H_2O \quad \rightarrow \quad NH_4^+ \quad + \quad OH^-$$

From the equation it should be easy to see that ammonia has one equivalent per mole.

Once the number of equivalents per mole for a compound has been determined, usually the next step is to calculate the equivalent mass. *The* **equivalent mass** *of a compound is the mass of that compound needed to produce one chemical equivalent.*

Some instructors may call this quantity the gram equivalent mass, equivalent weight or may simply use the units, grams/equivalent. The number of equivalents per mole provides an easy conversion to equivalent mass from molar mass.

$$\text{equivalent mass} = \frac{\text{molar mass}}{\text{equivalents/mole}}$$

It is important to note at this time that the equivalents per mole for a compound (and therefore also the equivalent mass) is determined by its behavior in a given chemical reaction. This doesn't present many problems in acid-base chemistry, but it does create some difficulty for oxidation-reduction reactions. Examples of this situation will be discussed later in this section.
The following calculation determins that 36.46 grams of HCl represents one equivalent in typical acid-base reactions. If the mass of HCl is greater than or less than 36.46 grams, obtain the number of equivalents by dividing the available mass by 36.46 g/eq. Notice that equivalents are not the same as equivalents per mole. The only limits on the number of equivalents of HCl are set by the ability to obtain different masses of this compound, but the number of equivalents per mole for HCl is established by the ionization equation. In order to change the eq/mol for HCl it would be necessary to change the ionization equation.

EXAMPLE 1. CALCULATION OF EQUIVALENTS

Calculate the number of equivalents in a 2.00 gram sample of (a) HCl and (b) $Ba(OH)_2$

(a) As seen above, the ionization of HCl produces one mole of hydrogen ions per mole of HCl or one eq/mol. Therefore, the equivalent mass is numerically equal to the molar mass.

$$\text{equivalent mass} = \frac{36.46 \text{ g/mol}}{1 \text{ eq/mol}} = 36.46 \text{ g/eq}$$

74

Divide the mass of sample by the equivalent mass to obtain equivalents

$$\text{eq of HCl} = \frac{2.00 \text{ g}}{36.46 \text{ g/eq}} = 0.0549 \text{ eq}$$

(b) The ionization equation

$$Ba(OH)_2 \rightarrow Ba^{2+} + 2 OH^-$$

indicates two equivalents per mole for this compound. Using the standard equation to calculate the equivalent mass

$$\text{equivalent mass} = \frac{171.3 \text{ g/mol}}{2 \text{ eq/mol}} = 85.66 \text{ g/eq}$$

As in the previous example, the sample mass is divided by the equivalent mass.

$$\text{eq of Ba(OH)}_2 = \frac{2.00 \text{ g}}{85.66 \text{ g/eq}} = 0.0233 \text{ eq}$$

Since most of the acid-base reactions that will be of concern occur in solution, it is necessary to create a concentration unit based on equivalents. The new unit, called Normality, is analogous to the molar concentration unit explained previously. Just as Molarity is defined as moles of solute per liter of solution, **Normality** *is defined as equivalents of solute per liter of solution.* The units of Normality are eq/liter, but it is common to use N as a symbol for this unit. Another way to think about Normality is that it is simply the number of moles per liter of hydrogen ion or hydroxide ion in a solution.

Notice that since the number of equivalents per mole is always equal to or greater than a value of one for acids and bases, one mole must always represent one or more equivalents. This means that Normality can be equal to or greater than Molarity but never less than the Molarity.

Another way to represent the relationship

between Normality and Molarity is

$$\text{Normality} = \text{Molarity} \times \frac{\text{equivalents}}{\text{mole}}$$

This equation is very useful for converting from one of these concentration units to another. To prove that it is correct, use dimensional analysis.

$$\frac{\text{equivalents}}{\text{liter}} = \frac{\text{moles}}{\text{liter}} \times \frac{\text{equivalents}}{\text{mole}}$$

Notice that when moles are canceled, the result is, indeed, an equality.

EXAMPLE 2. CONVERTING MOLARITY TO NORMALITY

Calculate the Normality of a 3.0 Molar solution of H_3PO_4.

Solution

First, write the ionization reaction for H_3PO_4 and determine the number of equivalents per mole.

$$H_3PO_4 \rightarrow 3 H^+ + PO_4^{3-}$$

There are three equivalents per mole for H_3PO_4.

Next use the relationship between Normality and Molarity developed above.

$$\text{Normality} = \text{Molarity} \times \frac{\text{equivalents}}{\text{mole}}$$

Substituting,

Normality = 3.0 Molar x 3 eq/mol = 9.0 N

Normality = 9.0 N

Thus a 3.0 Molar solution of H_3PO_4 has a Normality of 9.0 Normal in typical acid-base reactions.

SECTION 2. USING EQUIVALENTS AND NORMALITY IN ACID-BASE TITRATIONS

The preceding section explained that equivalents are a measure of the amount of acid or base in terms of the ability to produce or react with hydrogen ions. One equivalent of any acid will produce one mole of hydrogen ions, and one equivalent of any base will react with one mole of hydrogen ions. It follows that one equivalent of any acid is required to react completely with one equivalent of any base. This is the definition of the equivalence point for any acid-base titration. It can also be stated in the form of an equation. At the **equivalence point** of an acid-base titration,

Equivalents of acid = Equivalents of base

Chemists also often talk about the end point of a titration, and it is reasonable to ask how it is related to the equivalence point. The **end point** of a titration occurs when sufficient titrant has been added to cause a chemical indicator to change color. Since these chemical indicators don't change at exactly the point where the equivalents are equal, the end point is not exactly the same as the equivalence point. The two values are usually close enough, however, so normally the value of the end point and the equivalence point are presumed to be the same.

The statement that equivalents of acid equals equivalents of base is the starting point for most of the problems in this section. Almost all of the equations required for acid-base titrations can be obtained by substituting a few simple relationships into this relationship based on equivalents. For example, the equation

Normality = equivalents/liter.

can easily be rearranged to the form

Normality x Volume (in liters) = equivalents.

It has also been shown earlier that

equivalents = grams/equivalent mass.

Depending on what information is available, either of these can be substituted for equivalents of acid or base in the original equation. Most titration problems can be worked with some combination of these three equations.

Students often complain that there are too many equations to memorize when doing acid-base calculations with equivalents. There appear to be many different equations, but they are not all independent. Actually most of them are obtained by simple substitutions, such as those described in the previous paragraph. Once this is recognized, it's easy to design equations for each problem that is encountered. Thus, this is an important technique to master.

EXAMPLE 3. USING EQUIVALENTS IN AN ACID-BASE TITRATION

If 25.84 mL of 0.120 N HCl is required to titrate 100.0 mL of barium hydroxide to the equivalence point, what is the Normality of the $Ba(OH)_2$?

Solution

The first piece of important information is that the calculation deals with a titration at the equivalence point. Therefore, begin with the equation

eq of $Ba(OH)_2$ = eq of HCl

Since the Normality and the volume of the hydrochloric acid is given, it is reasonable to substitute eq = N x V on the hydrochloric acid side of the equation. The volume of the barium hydroxide is given and the Normality is the unknown, so also insert eq = N x V on that side of the equation.

78

The resulting equation is

$$N (Ba(OH)_2) \times V (Ba(OH)_2) = N (HCl) \times V (HCl)$$

Notice that strictly speaking both of these volumes should be in liters. In this case, since the same conversion factor will occur on both sides of the equality sign, it will cancel out. This means that the correct answer would result even if the volumes are not converted to liters. This is not always true. The next example is a case where failure to convert to liters will produce an incorrect answer. It's never wrong to convert all volumes to liters in these problems, so for students who are unsure, the best strategy is always to make the conversion.

Now, inserting the known values

$$N (Ba(OH)_2) \times 0.1000 \text{ L} = 0.120 \text{ N} \times 0.02584 \text{ L}$$

$$N (Ba(OH)_2) = \underline{0.0310 \text{ N}}$$

EXAMPLE 4. USING EQUIVALENTS WITH A SOLID REACTANT

How many grams of solid NaOH are required to neutralize 250.0 mL of 2.00 N H_3PO_4?

Solution

The language in this problem is somewhat different, since it doesn't specifically refer to the equivalence point. It does say that an acid and a base are being reacted until they neutralize each other, and this occurs at the equivalence point.
Thus, we can use the same relationship.

$$\text{eq of base (NaOH)} = \text{eq of acid } (H_3PO_4)$$

Since the volume and Normality of the H_3PO_4 are given, substitute eq = N x V on the right hand side of the equation. Neither Normality nor volume is available for the NaOH, and so a different relationship is needed. To find grams, use the relationship that eq = grams /equivalent mass. Is enough information provided to determine the equivalent mass for NaOH? Yes!

This completes the left hand side of the equation.

$$\frac{\text{grams (NaOH)}}{\text{equivalent mass (NaOH)}} = N \ (H_3PO_4) \ x \ V \ (H_3PO_4)$$

Now calculate the equivalent mass for NaOH. This compound has one equivalent per mole, so the equivalent mass is numerically the same as the molar mass, or 40.0 g/mol.

Notice that in this problem failure to convert the volume to liters will produce a large error in the answer. For this reason it is essential to convert the volume before substituting the values in the above equation.

$$\frac{\text{grams (NaOH)}}{40.00 \text{ g/eq}} = 2.00 \text{ N x } 0.250 \text{ L}$$

grams (NaOH) = 2.00 N x 0.250 L x 40.0 g/eq

grams (NaOH) = 20.0 grams

There is one problem type that may cause confusion when using the above method to solve titration problems. If the mass of the sample is given and the percentage of some component in that sample is the unknown, it's tempting to use the sample mass directly in the equivalents relationship. This is incorrect. The sample consists of at least two components, an acid or base that is being titrated, and some inert material that does not participate in the reaction. If this is the case, the mass provided cannot be substituted into the equivalents relationship, since it does not represent the amount of acid or base.

When a mass is given, always be sure to note whether this is the mass of the sample or the mass of a material contained in the sample.

80

EXAMPLE 5. CHEMICAL ANALYSIS USING EQUIVALENTS

Fumaric acid, $C_4H_4O_4$, is a substance essential for vegetable and animal respiration. It can produce two moles of ionizable hydrogen ions per mole of acid. Determine the weight percent of fumaric acid in a sample that is a mixture consisting only of fumaric acid and a solid, inert material if it requires 25.35 mL of 0.650 N NaOH to titrate a 2.010 gram sample of this mixture to the equivalence point?

Solution

This titration has gone to the equivalence point, so the first step is to write the equation

eq of fumaric acid = eq of sodium hydroxide

The Normality and the volume of the sodium hydroxide are given, and so it is obvious that N (NaOH) x V (NaOH) can be substituted for eq of NaOH on the right hand side of the equation. As was just noted, the 2.010 gram mass is not fumaric acid, but a mixture; however, the mass of fumaric acid in that mixture is needed.

Since the molar mass and the equivalents per mole are given for fumaric acid, calculate the equivalent mass of this substance. Then, on the left side of the equation, substitute grams divided by equivalent mass for equivalents of fumaric acid.

First, calculate the equivalent mass of fumaric acid

$$\text{equivalent mass} = \frac{\text{molar mass}}{\text{eq/mol}}$$

$$\text{equivalent mass} = \frac{116.08 \text{ g/mol}}{2 \text{ eq/mol}} = 58.04 \text{ g/eq}$$

Next, make the appropriate substitutions in the equivalents relationship

$$\frac{\text{gram of fumaric acid}}{\text{eq mass of fumaric acid}} = N\ (NaOH)\ \times\ V\ (NaOH)$$

Convert the volume to liters, insert the known values and solve for grams of fumaric acid.

$$\frac{\text{grams of fumaric acid}}{58.04\ g/eq} = 0.650\ N\ \times\ 0.02535\ L$$

grams of fumaric acid = 0.9564 grams

Finally, determine the weight percent of fumaric acid in the sample.

$$\text{weight percent} = \frac{\text{grams of fumaric acid} \times 100}{\text{grams of sample}}$$

$$\text{weight percent} = \frac{0.9564\ g\ \times\ 100}{2.010\ g}$$

weight percent = 47.6%

Notice that in each case, the essential relationship has been the observation that the equivalents of acid must equal the equivalents of base. This must be true at the equivalence point of any titration according to the definition.

In some problems there may be more than one acid or base. This method is sometimes used intentionally and called a back titration. This situation does not, however, require any change in the solution method. At the equivalence point, the total number of equivalents of all of the acids present still must be identical with the total number of equivalents of all of the bases present. If there are two different acids used to titrate a base to the equivalence point, the equivalents relationship becomes

eq of acid 1 + eq of acid 2 = eq of base

and the same equations are substituted for equivalents as in the previous examples.

SECTION 3. USING EQUIVALENTS FOR OXIDATION-REDUCTION REACTIONS

Chemical equivalents can also be used for oxidation-reduction reactions, but some change is needed, since hydrogen ions are no longer the best measure of chemical reactivity. In the previous study of oxidation-reduction processes, it was pointed out that the transfer of electrons was the essential component in these reactions. This suggests that a different definition of the chemical equivalent is appropriate.

For oxidation-reduction reactions, a **chemical equivalent** *is defined as the amount of substance that will react with or produce one mole of electrons.*

It is rarely possible to determine equivalents per mole for oxidizing agents or reducing agents by inspection of either the formula of the substance or the balanced equation. *Equivalents per mole for oxidizing agents or reducing agents are determined from the balanced half-reactions.* Unlike acid-base problems, there are few shortcuts here; however, once it is recognized that the equivalents per mole cannot be determined by inspection of the formula, the process is not really too difficult.

It has already been pointed out in Chapter 5 of the textbook that *loss of electrons is oxidation and gain of electrons is reduction.* Also, the substance oxidized in a chemical reaction is the *reducing agent*, and the substance reduced is the *oxidizing agent*.

In the left-hand column below are a list of balanced half-reactions. The right-hand column lists the number of eq/mol for the reactant in each half-reaction.

half-reaction	eq/mol
$Cu^{2+} + 2\ e^- \rightarrow Cu$	2
$Fe^{2+} \rightarrow Fe^{3+} + e^-$	1
$MnO_4^- + 8\ H^+ + 5\ e^- \rightarrow Mn^{2+} + 4\ H_2O$	5
$2\ Cl^- \rightarrow Cl_2 + 2\ e^-$	1

Notice that in the last reaction, two moles of electrons are produced, but this also requires two moles of chloride ions. Thus there are <u>two</u> equivalents for <u>two</u> moles of chloride, or one equivalent per mole.

Aside from the method for determining equivalents per mole, most of the other techniques necessary to apply equivalents to oxidation-reduction reactions are very similar to those previously learned. Terms such as Normality, equivalent mass, and equivalence point still have the same meaning. The only important difference is the change in the method of determining equivalents per mole.

EXAMPLE 6. USING EQUIVALENTS FOR OXIDATION-REDUCTION REACTIONS

Calculate the Normality of a certain potassium permanganate, $KMnO_4$, solution if 26.90 mL of this solution is required to titrate 1.00 gram of iron(II) to the equivalence point in acidic solution. The skeleton equation (not balanced) is

$$Fe^{2+} \quad + \quad MnO_4^- \quad \rightarrow \quad Fe^{3+} \quad + \quad Mn^{2+}$$

Now calculate the Molarity from the Normality.

Solution

As was the case in acid-base problems, the first step is to recognize that

eq of oxidizing agent = eq of reducing agent

Iron is the reducing agent, since it is oxidized from +2 to +3. Permanganate ion is the oxidizing agent, since manganese is reduced from +5 to +2.

Next determine what to substitute into the equivalents relationship. The mass of iron is given, and the equivalent mass of iron can be determined from the information given. It is reasonable, then, to substitute grams divided by equivalent mass for equivalents of reducing agent. The volume of permanganate is given, and the Normality is the unknown, so substitute N x V for equivalents of oxidizing agent. (Don't forget that the volume must be in liters.)

Before making these substitutions, determine the equivalent mass for iron(II). From the skeleton equation, iron(II) is oxidized to iron(III), and the balanced half-reaction is easy to write:

$$Fe^{2+} \quad \rightarrow \quad Fe^{3+} \quad + \quad e^-$$

This indicates that there is one equivalent per mole for iron(II). Since iron(II) is an ion, not a molecule, the equivalent mass is obtained by dividing the atomic weight by the equivalents per mole

$$\text{equivalent mass} \quad = \quad \frac{\text{atomic weight}}{\text{eq/mole}}$$

$$\text{equivalent mass} \quad = \quad \frac{55.85 \text{ g/mol}}{1 \text{ eq/mole}}$$

$$\underline{\text{equivalent mass} \quad = \quad 55.85 \text{ g/eq}}$$

Now substitute into the equivalents relationship.

$$N \text{ (MnO}_4^-) \text{ x V (MnO}_4^-) \quad = \quad \frac{\text{grams Fe}^{2+}}{\text{eq mass of Fe}^{2+}}$$

$$N \text{ (MnO}_4^-) \text{ x .0269 L} \quad = \quad \frac{1.00 \text{ g}}{55.85 \text{ g/eq}}$$

$$\underline{N \text{ (MnO}_4^-) \quad = \quad 0.666 \text{ N}}$$

There are several different ways to determine the Molarity of this solution but probably the simplest is to use the equation

Normality = Molarity x eq/mol

To determine the eq/mol for permanganate, it's

necessary to use the balanced half-reaction. From the skeleton reaction given, the basic components are

$$MnO_4^- \rightarrow Mn^{2+}$$

Balancing this equation produces

$$MnO_4^- + 8\ H^+ + 5\ e^- \rightarrow Mn^{2+} + 4\ H_2O$$

This indicates that there are five eq/mol for MnO_4^-. Substituting this value and the Normality into the original equation

0.666 N = Molarity x 5 eq/mol

Molarity = 0.133 M

In the previous problem, the reactants were all ionic. In many cases, the amount of the reactants is given in terms of the compound which serves as a source of the ion. For instance, what if the preceding problem had been stated in terms of 1.00 gram of iron(II) chloride rather than 1.00 gram of iron(II)? The same method would still be used to determine eq/mol, but the equivalent mass would change. Instead of obtaining the equivalent mass by dividing the atomic weight of iron by 1 eq/mol, divide the molar mass of iron(II) chloride by 1 eq/mol. Otherwise, the set-up is the same.

It is important to realize that the equivalent mass of a species depends on the specific reaction in which it is used. For example, suppose that the reaction described in Example 6 were done in basic solution rather than acidic solution. The product of the permanganate reduction would then be manganese(IV) oxide rather than manganese(II) ion. The permanganate half-reaction is now

$$MnO_4^- + 2\ H_2O + 3\ e^- \rightarrow MnO_2 + 4\ OH^-$$

It is clear that permanganate now has three equivalents per mole. This means that even if the permanganate for the titration is taken from exactly the same stock solution, it will have a different Normality, because the equivalents per mole will be different.

To define the Normality of a solution, it is **essential** to know the equation for the reaction in which the solution is to be used. If a change in the conditions alters that equation, this will change the Normality of the solution. This can be accomplished by converting the old Normality to Molarity, then determining the new Normality based on that Molarity. Changing the reaction doesn't change the Molarity, so it provides a constant value for comparison with the Normality.

EXAMPLE 8. CALCULATING A NEW NORMALITY WHEN EQUIVALENTS/MOLE CHANGES

The Normality of a certain solution of potassium permanganate is determined to be 6.0 Normal in acidic solution where the half reaction is

$$MnO_4^- + 8 H^+ + 5 e^- \rightarrow Mn^{2+} + 4 H_2O$$

Calculate the Normality of that same solution when the potassium permanganate is to be used in basic solution where the half-reaction is

$$MnO_4^- + 2 H_2O + 3 e^- \rightarrow MnO_2 + 4 OH^-$$

Solution

First, convert the original Normality to Molarity. From the half-reaction, it is clear that in acidic solution there are five equivalents per mole for permanganate. Substituting the available information into the equation that relates Normality and Molarity

Normality = Molarity x eq/mol

6.0 N = Molarity x 5 eq/mol

Molarity = 1.2 M

Next, examination of the half-reaction for basic conditions indicates that there are now three

eq/mol.   Use the same relationship as before to
convert the Molarity to the new Normality.

Normality = Molarity x eq/mol

Normality = 1.2 M x 3 eq/mol

Normality = 3.6 N

It has been shown that it's slightly more
difficult to set up titration problems using
Normality that it would be using Molarity.
Determining equivalents per mole is not much harder
than using the balanced equation to find a
stoichiometric relationship, but using equivalents
does require some new concepts. On the other hand,
when doing the same titration over and over again,
it is easier to use Normality,   especially if
several reactants must be related to each other.
*Regardless of whether Normality or Molarity is used
for these problems, it is important to recognize
that both methods represent the same basic chemical
concepts.*

STUDY QUESTIONS FOR CHEMICAL EQUIVALENTS

1. Barium hydroxide reacts with HCl according to
the balanced equation

$Ba(OH)_2(aq) + 2 HCl(aq) \rightarrow BaCl_2(aq) + 2 H_2O(\ell)$

(a) If 1.7 grams of barium hydroxide are dissolved
in enough water to produce 125 mL of solution, what
is the Molarity of the resulting solution?   (b)
What is the Normality of the solution? (c) What is
the molar concentration of OH^- in this solution?

2. Succinic acid, $C_4H_6O_4$, loses two hydrogen ions on
reaction with a base. If 2.50 grams of this acid
is dissolved in enough water to produce 255 mL of
solution, (a) what is the Normality of the
resulting solution?   (b) What is the molar
concentration of H^+ in this solution?

3. Calculate the Normality of a sulfuric acid solution prepared by dissolving 25.0 grams of H_2SO_4 in enough water to produce 150. mL of solution. What is the Molarity of this solution? What is the concentration of H^+ in this solution?

4. (a) How many equivalents of acid are contained in 300.0 mL of 0.15 N oxalic acid, $H_2C_2O_4$? (b) If this solution is diluted to a volume of 500.0 mL, what is the new Normality? (c) Did the number of equivalents of acid change when the solution was diluted?

5. (a) How many milliliters of 0.250 N HCl would be required to neutralize completely 2.50 grams of NaOH? (b) How many milliliters of 0.250 N H_2SO_4 would be require to neutralize the same mass of NaOH?

6. How many milliliters of 0.110 N HCl will be required to titrate 5.00 grams of $Ba(OH)_2$ to the equivalence point?

7. Calculate the equivalent mass of a certain unknown acid if 31.0 mL of 0.132 N NaOH is require to titrate a 0.500 gram sample of the unknown acid to the equivalence point.

8. A 1.00 gram sample of citric acid, $C_6H_8O_7$, requires 31.23 mL of 0.500 N NaOH for titration to the equivalence point. (a) What is the equivalent mass of citric acid? (b) What is the number of equivalents per mole for citric acid?

9. What is the percent by weight of acetic acid in a vinegar sample if it requires 30.24 mL of 0.210 N KOH to neutralize the acid in 10.0 mL of vinegar. You may assume all of the acidity is due to acetic acid, $HC_2H_3O_2$, and that the density of the vinegar is 1.00 g/mL.

10. Suppose that you are attempting to determine the equivalent mass of a substance commonly used in antacid tablets. You dissolve 0.210 grams of the pure material (which is a base) in 25.0 mL of 0.400

N HCl and then titrate the resulting mixture with 0.110 N NaOH. If 18.2 mL of NaOH are required to reach the end point, what is the equivalent mass of the antacid?

11. A solution is prepared by mixing 31.21 mL of 0.100 N HCl with 98.53 mL of 0.500 N H_2SO_4 and then adding 50.0 mL of 1.002 N $Ca(OH)_2$ to the resulting mixture. (a) Is the resulting solution acidic or basic? (b) How many milliliters of 0.300 N acid or base (as appropriate) must be added to the solution to make it exactly neutral?

12. The skeleton equation (not balanced) for the reduction of iron(II) with tin(II) is as follows:

$$Fe^{3+} + Sn^{2+} \rightarrow Fe^{2+} + Sn^{4+}$$

Answer the following questions based on this equation. (a) What is the equivalent mass of tin(II) chloride? (b) How many grams of $SnCl_2$ must be dissolved in 2.00 liters of water to produce a 0.200 N solution of tin(II)? (c) How many equivalents of iron(III) are present in a 5.00 gram sample of iron(III) chloride? (d) How many milliliters of 0.200 N tin(II) solution will be required to exactly reduce 5.00 grams of iron(III) chloride to iron(II) chloride?

13. The equation for the reaction of hydrogen sulfide with iodine is

$$H_2S + I_2 \rightarrow 2 I^- + 2 H^+ + S$$

According to this reaction, (a) How many grams of I_2 are contained in 0.750 liters of a 0.330 N iodine solution? (b) How many milliliters of 0.330 N iodine solution will be required to react completely with 0.225 grams of H_2S?

14. The skeleton equation (not balanced) for the reaction of iron(II) with permanganate ion in acidic solution is as follows:

$$MnO_4^- + Fe^{2+} \rightarrow Mn^{2+} + Fe^{3+}$$

90

Answer the following questions based on this equation.
(a) Calculate the number of equivalents in 25.0 mL of 0.110 N MnO_4^-. (b) How many grams of $KMnO_4$ must be dissolved in 250. mL of water to produce a 0.110 N solution? (c) Calculate the number of equivalents in 2.00 grams of $FeCl_2$. (d) How many milliliters of 0.110 N $KMnO_4$ will be required to titrate a 2.00 gram sample of pure iron(II) chloride to the equivalence point?

15. When permanganate ion acts as an oxidizing agent in basic solution, the product is MnO_2, as shown in the balanced half-reaction,

$$MnO_4^- + 2 H_2O + 3 e^- \rightarrow MnO_2 + 4 OH^-$$

The solution of potassium permanganate described in the preceding question had a Normality of 0.110 when it was used in acidic reactions. What would be the Normality of this same permanganate when reacting in basic solution to form MnO_2, as shown in the above half-reaction?

16. The compound $Fe(NH_4)_2(SO_4)_2 \cdot 6H_2O$ (molar mass = 392.19 g/mol) is sometimes used to standardize permanganate solutions. If 32.45 mL of aqueous $KMnO_4$ is required to titrate the iron(II) ion in a 1.050 gram sample of this compound to the equivalence point, what is the Normality of the $KMnO_4$? The balanced equation for the reaction is

$$MnO_4^- + 5 Fe^{2+} + 8 H^+ \rightarrow Mn^{2+} + 5 Fe^{3+} + 4 H_2O$$

17. A sample of iron ore weighing 2.010 grams was dissolved in acid, and the iron was all reduced to soluble iron(II). What is the weight percent of iron in the ore sample if 34.7 mL of 0.142 N $KMnO_4$ is required to titrate the iron(II) to the equivalence point? The balanced equation for this reaction is given in the previous problem.

18. Zinc oxide is used as a pigment in white paints. It is basic and will react with acids according to the equation

$$ZnO + 2H^+ \rightarrow H_2O + Zn^{2+}$$

A 2.820 gram sample of technical grade (i.e. not pure) zinc oxide is dissolved in 50.0 mL of 1.00 N sulfuric acid and not all of the acid is neutralized. The resulting solution is then titrated with 0.137 N KOH, and 3.30 mL of KOH is required to reach the equivalence point. What is the weight percent of ZnO in the original sample? (Assume that the sample contains no basic material other than the ZnO.)

19. A solution of nitric acid, HNO_3, is prepared to be 4.00 N when acting as an acid. Later on it is decided to use this solution in a redox reaction that involves the reduction of nitric acid to NO. How many mL of water must be added to a 75.0 mL sample of the original nitric acid solution in order to make it 3.00 N when acting as an oxidizing agent?

STUDY QUESTION ANSWERS

1. a. 0.079 M b. 0.160 N c. 0.160 M
2. a. 0.166 N b. 0.166 M
3. 3.40 N, 1.70 M, 3.40 M
4. a. 0.045 equivalents b. 0.090 N c. no
5. a. 250 mL b. 250 mL
6. 531 milliliters
7. 122 g/eq
8. a. 64.0 g/eq b. 3 eq/mole
9. 3.81%
10. 26.3 g/eq
11. a. acidic b. 7.6 mL
12. a. 94.80 g/eq b. 37.9 g
    c. 0.0308 eq   d. 154 mL
13. a. 31.4 g b. 40.0 mL
14. a. 0.00275 eq b. 0.869 g
    c. 0.0158 d. 143 mL
15. 0.0660 N
16. 0.08250 N
17. 13.7% Fe
18. 71.5% (that is, 2.02 g of ZnO)
19. 225 mL

SPECIAL SECTION II
STUDY MATERIAL FOR
NORMALITY AND EQUIVALENTS

LEARNING GOALS

1. Be able to determine the number of equivalents per mole and the equivalent mass for common acids and bases. From this information, determine how many equivalents are contained in a given acid or base sample. (Sec. I.1)

2. Understand how to use the relationship that states equivalents of acid = equivalents of base to solve problems involving neutralization reactions or acid-base titrations at the equivalence point. (Sec. I.2)

3. Given the reactants and products in a half reaction, be able to determine the number of equivalents per mole and the equivalent mass for oxidizing agents and reducing agents. Based on this information, be able to determine how many equivalents are contained in a sample of oxidizing or reducing agent. (Sec. I.3)

4. Using the equivalents of oxidizing agent = equivalents of reducing agent relationship, be able to solve problems involving oxidation-reduction titrations at the equivalence point. (Sec. I.3)

5. For either acid-base or oxidation-reduction reactions, be able to calculate the new Normality from an old Normality, if given the chemical equations used to define these two quantities. (Sec. I.3)

CONCEPT TEST

1. For acid-base reactions, one equivalent of a substance is defined to be the amount of that substance that will combine with or produce one mole of _____.

2. For oxidation-reduction reactions, one equivalent of a substance is defined as the amount of that substance that will combine with or produce one mole of _____ .

3. Indicate the number of equivalents per mole for each of the following acids and bases. (Assume complete neutralization.)

a. H_2SO_4 _____   b. NaOH _____

c. H_3PO_4 _____   d. $Ca(OH)_2$ _____

4. Determine equivalents per mole for the reactant underlined in each balanced half reaction.

half reaction	eq/mol
a. $\underline{Zn} \rightarrow Zn^{2+} + 2\ e^{-}$	_____
b. $\underline{Br_2} + 2\ e^{-} \rightarrow 2\ Br^{-}$	_____
c. $\underline{NO_3^{-}} + 4\ H^{+} + e^{-} \rightarrow NO + 2\ H_2O$	_____

5. The _____ of a compound is the mass of that compound needed to produce one chemical equivalent.

6. _____ is defined as equivalents of solute per liter of solution.

7. The equivalence point of an acid-base titration occurs when the equivalents of acid equals the

_____ .

8. The _____ of an acid-base titration occurs when sufficient titrant has been added to cause a chemical indicator to change color.

STUDY HINTS

1. As you will see, the titration problems discussed in this section are always at the equivalence point, which allows us to use the equivalents relationship. You may well wonder what

to do if you find a problem in which the titration is not at the equivalence point. This is more complicated, and problems of this type will not be encountered until later in the course.

2. Be sure to understand the difference between equivalents and equivalents per mole. To draw a rough analogy, equivalents per mole is comparable to the cost of bread per loaf and equivalents can be compared with how much money you have. If bread costs 80 cents per loaf, that is a fixed amount, but doesn't determine how much money you have. It does mean that if you have $1.60, you can buy two loaves of bread. If a certain reactant is found to have two equivalents per mole, this doesn't mean you always must have two equivalents, but it does mean that if you have one mole you will have two equivalents. For a given chemical in a given reaction, the number of equivalents per mole is fixed, but you may have any number of equivalents depending on how many grams of substance are available.

3. Be sure to read the problems carefully to determine whether you are doing calculations that deal with an ion or a complete compound. For example, a problem can ask what is the percentage of iron(II) in a sample or what is the percentage of iron(II) chloride in a sample. It is easy to overlook this distinction in the heat of a test.

PRACTICE PROBLEMS

1. a. Calculate the Normality of a nitric acid solution prepared by dissolving 50.0 grams of HNO_3 in enough water to produce 235 ml of solution. You may assume that the nitric acid will be used in a simple acid-base reaction. b. What is the Molarity of this solution? c. What is the concentration of H^+ in this solution?

2. Calculate the number of equivalents in a 125 gram sample of phosphoric acid, H_3PO_4. When phosphoric acid reacts with a strong base, such as NaOH, three moles of hydrogen ions can react per mole of acid.

3. How many milliliters of 0.110 N H_2SO_4 will be required to titrate 5.00 grams of KOH to the equivalence point?

4. What is the equivalent mass of a certain unknown acid if 27.3 ml of 0.210 N KOH is required to titrate a 0.361 gram sample of the unknown acid to the end point.

5. Calculate the Normality of a solution of a sodium thiosulfate solution if 30.0 ml of this solution is required to titrate 1.43 grams of iodine according to the equation

$$I_2 + 2\ Na_2S_2O_3 \rightarrow 2\ NaI + Na_2S_4O_6$$

6. The skeleton equation (not balanced) for the reduction of iron(II) with permanganate in acidic solution is

$$Fe^{2+} + MnO_4^- \rightarrow Mn^{2+} + Fe^{3+}$$

Answer the following questions based on this equation.
(a) What is the equivalent mass of potassium permanganate?
(b) If you wish to produce a 0.200 N solution of potassium permanganate, how many grams of $KMnO_4$ must be dissolved in 2.00 liters of water?
(c) How many equivalents of iron(II) are present in a 5.00 gram sample of iron(II) chloride?
(d) How many milliliters of 0.200 N potassium permanganate solution will be required to exactly oxidize 5.00 grams of iron(II) chloride to iron(III) chloride?

7. The equivalent mass of benzoic acid is 122.1 g/eq. An unknown sample having a mass of 0.800 grams is a mixture of benzoic acid and inert material. If 24.7 ml of 0.206 N KOH is required to titrate this sample to the equivalence point, (a) how many grams of benzoic acid were in the original sample, and (b) what is the percent by weight of benzoic acid in the original sample?

PRACTICE PROBLEM SOLUTIONS

1. a. If we knew the equivalent mass of nitric acid, the Normality could be calculated using the equation

$$\text{Normality} = \frac{\text{g nitric acid/ eq mass nitric acid}}{\text{volume of solution (L)}}$$

In an acid-base reaction, there is one equivalent per mole for nitric acid, and so the equivalent mass is obtained by

$$\text{equivalent mass} = \frac{\text{molar mass of nitric acid}}{\text{eq/mol for nitric acid}}$$

$$\text{equivalent mass} = \frac{63.02 \text{ g/mol}}{1 \text{ eq/mol}}$$

equivalent mass = 63.02 g/eq

Substituting into the equation for Normality

$$\text{Normality} = \frac{50.0 \text{ g /63.02 g/eq}}{0.235 \text{ L}}$$

Normality = 3.38 N

b. Use the equation that relates Normality and Molarity

Normality = Molarity x eq/mol

$$\text{Molarity} = \frac{\text{Normality}}{\text{eq/mol}} = \frac{3.38 \text{ N}}{1 \text{ eq/mol}}$$

Molarity = 3.38 M

c. The hydrogen ion concentration is numerically equal to the Normality expressed in molar concentration units, and so

H^+ concentration = 3.38 M

2. The number of equivalents per mole is given as three in the problem. To find equivalents, we can use the equation

$$\text{equivalents} = \frac{\text{mass}}{\text{equivalent mass}}$$

The equivalent mass of H_3PO_4 is given by the equation

$$\text{equivalent mass} = \frac{\text{molar mass}}{\text{eq/mol}} = \frac{98.00 \text{ g/mol}}{3 \text{ eq/mol}}$$

equivalent mass = 32.67 g/eq

Substituting the equivalent mass into the equation for equivalents

$$\text{equivalents} = \frac{125 \text{ g}}{32.67 \text{ g/eq}}$$

equivalents = 3.83 eq

3. Since the reaction is at the equivalence point

$$\text{equivalents of } H_2SO_4 = \text{equivalents of KOH}$$

For sulfuric acid, we know the Normality and wish to find the volume, so we can substitute

$$\text{eq of } H_2SO_4 = \text{Normality of } H_2SO_4 \text{ x volume of } H_2SO_4$$

For KOH, we know the mass, and can readily determine the equivalent mass, since there is 1 eq/mole for KOH.

$$\text{eq mass of KOH} = \frac{\text{molar mass}}{\text{eq/mol}} = \frac{56.11 \text{ g/mol}}{1 \text{ eq/mol}}$$

eq mass of KOH = 56.11 g/eq

$$\text{equivalents of KOH} = \frac{\text{grams of KOH}}{\text{equivalent mass of KOH}}$$

The equivalents relationship has now become

$$\text{Normality } (H_2SO_4) \times \text{volume } (H_2SO_4) = \frac{\text{grams (KOH)}}{\text{eq mass (KOH)}}$$

Substituting the known values

$$0.110 \text{ N} \times V(H_2SO_4) = \frac{5.00 \text{ g}}{56.11 \text{ g/eq}}$$

Solving for the volume of sulfuric acid (remember it will be in liters).

$$V(H_2SO_4) = 0.810 \text{ L}$$

$$V(H_2SO_4) = 0.810 \text{ L} \times 1000 \text{ mL/L}$$

$$V(H_2SO_4) = \underline{8.10 \times 10^2 \text{ ml}}$$

4. Since the titration is at the end point, begin with the relationship

equivalents of base = equivalents of acid

Examining the data suggests we should substitute N x V for equivalents of base and grams/equivalent mass for equivalents of the acid. The equivalents relationship now becomes

$$\text{N(base)} \times \text{V(base)} = \frac{\text{grams (acid)}}{\text{equivalent mass (acid)}}$$

Substitute the available information and solve for the equivalent mass. (Don't forget you must convert the volume to liters!)

$$0.210 \text{ N} \times 0.0273 \text{ L} = \frac{0.361 \text{ g}}{\text{eq mass (acid)}}$$

$$\underline{\text{equivalent mass (acid)} = 63.0 \text{ g/eq}}$$

5. If the iodine is to be completely titrated by the thiosulfate, this oxidation-reduction titration must be at the equivalence point, allowing us to write the relationship

eq (oxidizing agent) = eq (reducing agent)

Sodium thiosulfate is the reducing agent. We are given the volume and wish to find the Normality of this reagent, so it's easy to see that we should substitute N x V for equivalents of reducing agent.

Given the mass of iodine, we could substitute grams/equivalent mass = equivalents into the equivalents relationship. To find the equivalent mass of iodine, we must write the half reaction (notice that the NaI ionizes to form iodide ions).

$$2e + I_2 \rightarrow 2 I^-$$

Which indicates that there are 2 eq/mole for I_2.

Determine the equivalent mass using the equation

$$eq\ mass = \frac{molar\ mass}{eq/mole} = \frac{253.8\ g/mol}{2\ eq/mol} = 126.9\ g/eq$$

Now the specific equivalents relationship for this problem is

$$\frac{grams\ (I_2)}{eq\ mass\ (I_2)} = N\ (S_2O_3^{2-})\ x\ V\ (S_2O_3^{2-})$$

Now substitute the values given (but be sure to change the volume to liters) and solve.

$$\frac{1.43\ g}{126.9\ g/eq} = N\ (S_2O_3^{2-})\ x\ 0.0300\ L$$

$$\underline{N\ (S_2O_3^{2-}) = 0.376\ N}$$

6. a. First, write and balance the permanganate half reaction to determine the equivalents per mole

$$8H^+ + 5e + MnO_4^- \rightarrow Mn^{2+} + 4\ H_2O$$

There are 5 eq/mol for permanganate.
Next, determine the equivalent mass

$$eq\ mass = \frac{molar\ mass}{eq/mol} = \frac{158.04\ g/mol}{5\ eq/mol}$$

eq mass = 31.61 g/eq

Since the mass of potassium permanganate is indicated, use the molar mass of $KMnO_4$, not just the MnO_4^- ion.

b. Solve the equation for the number of equivalents

eq = N x V = 0.200 N x 2.00 L = 0.400 eq

and then multiply this number of equivalents times the equivalent mass to determine the number of grams needed.

grams = eq x eq mass = 31.61 g/eq x 0.400 eq

grams = 12.6 g $KMnO_4$

c. From the information given, we can write and balance the half reaction for iron

$$Fe^{2+} \rightarrow Fe^{3+} + e$$

which tells us that there is one equivalent per mole for iron(II) in this reaction.

Next, find the equivalent mass for iron(II) chloride (not just for the iron!).

eq mass ($FeCl_2$) = $\dfrac{\text{molar mass}}{\text{eq/mole}}$ = $\dfrac{126.7 \text{ g/mol}}{1 \text{ eq/mol}}$

equivalent mass ($FeCl_2$) = 126.7 g/eq

To find the number of equivalents, divide the grams by the equivalent mass

equivalents = $\dfrac{5.00 \text{ g}}{126.75 \text{ g/eq}}$

equivalents = 0.0394 eq

d. Since the titration is at the equivalence point,

equivalents of MnO_4^- = equivalents of $FeCl_2$

102

We can substitute N x V for equivalents of permanganate, and the equivalents of $FeCl_2$ were obtained in the previous section.

$$N\ (MnO_4^-)\ x\ V\ (MnO_4^-) = \text{equivalents of } FeCl_2$$

Substituting the values given

$$0.200\ N\ x\ V\ (MnO_4^-)\quad = 0.0394\ eq$$

$$V\ (MnO_4^-) = 0.197\ L$$

but the volume is requested in milliliters

$$V\ (MnO_4^-) = 0.197\ L\ x\ 1000\ mL/L$$

$$\underline{V\ (MnO_4^-) = \underline{197\ ml}}$$

7. a. As usual, begin with the equivalents relationship

equivalents of KOH = equivalents of benzoic acid

Substitute V x N for equivalents of KOH and grams/equivalent mass for equivalents of benzoic acid. Notice that the mass given is <u>not</u> the mass of pure benzoic acid!

$$V\ (KOH)\ x\ N\ (KOH) = \frac{\text{grams (benzoic acid)}}{\text{equivalent mass (benzoic acid)}}$$

$$0.0247\ L\ x\ 0.206\ N = \frac{\text{grams (benzoic acid)}}{122.1\ g/eq}$$

$$\underline{\text{grams (benzoic acid)} = 0.621\ g}$$

b. Now find the percent of the sample that is benzoic acid

$$\text{percent acid} = \frac{\text{grams of acid x 100}}{\text{grams of sample}}$$

$$\text{percent acid} = \frac{0.621\ g\ x\ 100}{0.800\ g}$$

PRACTICE TEST (40 Minutes)

1. Indicate the number of equivalents per mole for each of the following acids and bases. (Assume complete neutralization.)

a. H_2CO_3 _____   b. HNO_3 _____

c. NH_3 _____   d. $Ba(OH)_2$ _____

2. For each section below, the reactant and product are indicated for a skeleton oxidation-reduction half reaction. In each case write the balanced half reaction (in acidic solution) and indicate the equivalents per mole for the reactant.

<u>half reaction</u>   <u>eq/mol</u>

a. Aluminum(III) is converted to   _____
   aluminum metal.

b. Dichromate ion ($Cr_2O_7^{2-}$) is   _____
   converted to chromium(III).

c. Oxalate ion ($C_2O_4^{2-}$) is   _____
   is converted to carbon dioxide.

3. (a) How many equivalents of base are contained in 250.0 ml of 0.231 N calcium hydroxide, $Ca(OH)_2$? (b) If you dilute this solution to a volume of 500. ml, what is the new Normality? (c) Did the number of equivalents of base change when the solution was diluted?

4. Malic acid, $C_4H_6O_5$, which is commonly found in apples and other fruits, normally loses two hydrogen ions on reaction with a base. (a) What is the equivalent mass of this acid? (b) If 1.75 grams of malic acid is used in a reaction, how many equivalents of acid are used? (c) If the 1.75 gram sample of malic acid is dissolved in enough water to make 150. ml of solution, what is the Normality and Molarity of the solution?

5. A certain chromate solution (CrO_4^{2-}) is 3.0 N when it forms chromium(III) hydroxide in basic solution. Write the balanced half reaction and determine the Molarity of the chromate solution.

6. (a) How many milliliters of 0.250 N KOH would be required to completely neutralize 2.50 grams of oxalic acid, $H_2C_2O_4$? (Oxalic acid has two ionizable hydrogens.) (b) How many milliliters of 0.250 N $Ba(OH)_2$ would be require to neutralize the same mass of oxalic acid?

7. What is the Normality of a Sn^{2+} if it contains 1.18 grams of tin(II) per liter of solution and the tin will be used as a reducing agent according to the half reaction

$$Sn^{2+} \rightarrow Sn^{4+} + 2e^-$$

8. A sample of iron ore weighing 0.800 grams is dissolved in acid and all of the iron in the sample is reduced to iron(II). The sample is then titrated in acid solution with 0.100 Normal $KMnO_4$ according to the unbalanced equation

$$Fe^{2+} + MnO_4^- \rightarrow Mn^{2+} + Fe^{3+}$$

If exactly 28.0 ml of potassium permanganate is required to reach the end point of this titration, what is the percentage of iron in the original sample?

CONCEPT TEST ANSWERS

1. hydrogen ions (or protons)
2. electrons
3. a. two b. one c. three d. two
4. a. two b. two c. one
5. equivalent mass   6. Normality
7. equivalents of base   8. end point

PRACTICE TEST ANSWERS

1. a. two   b. one   c. one   d. two

2.   <u>half reaction</u>   <u>eq/mol</u>

a. $3 e^- + Al^{3+} \rightarrow Al$ 3

b. $6 e^- + 14 H^+ + Cr_2O_7^{2-}$
                    $\rightarrow 2 Cr^{3+} + 7 H_2O$ 6

c. $C_2O_4^{2-} \rightarrow 2 CO_2 + 2e^-$ 2

3. a. 0.0578 equivalents b. 0.116 c. no

4. a. 67.1 g/eq b. 0.0261 eq
   c. 0.174 N, 0.0870 M

5. $3e^- + 4 H_2O + CrO_4^{2-} \rightarrow Cr(OH)_3 + 5 OH^-$

   Molarity = 1.0 M

6. a. 222 ml b. 222 ml
   (Notice the amounts <u>must be the same</u>!)

7. 0.0199 N   8. 19.5%

CHAPTER 6
ENERGY AND
CHEMICAL REACTIONS

CHAPTER OVERVIEW

The two important components of chemical reactions are matter and energy. While we spend a great deal of time discussing how matter participates in chemical reactions, energy sometimes seems to be less important. That isn't true, since the study of energy changes provides a valuable indication of whether or not a reaction is spontaneous. This chapter looks at the role of energy in chemical reactions.

Before looking at chemical reactions, it is necessary to begin with the relationship between heat and temperature. This requires an understanding of specific heat capacity, and the heats involved in state changes.

There are two mathematical equations which are especially significant when studying chemical reactions. One is the first law of thermodynamics, which states that energy is conserved in chemical reactions. The second mathematical relationship is Hess's law, which uses the fact that enthalpy is a state function to allow us to calculate the enthalpy change for an overall process from the enthalpy values for the constituent reactions.

All of these concepts are brought together by the topic of calorimetry, which is the method used to measure the heat involved in a chemical reaction.

LEARNING GOALS

1. Recognize the various forms of energy and understand how the transformations from one form to another are governed by the conservation of energy principle. Also be able to convert energy values from one unit to another, for example, from calories to joules and from joules to calories. (Sec. 6.1)

2. Adding or removing heat causes a substance to change temperature and/or state. If no state change is involved, the specific heat equation

$$q = c \times m \times \Delta t$$

(or the molar heat capacity equation) is used. (Sec. 6.2)

3. The equation for state changes is

$$q = \Delta H \times m$$

where ΔH may be the value for fusion or vaporization. Be able to do problems that use either this equation or the equation in the previous learning objective, or a combination of both equations. Your instructor may also ask you to memorize the values of the heat of vaporization and fusion for water. (Sec. 6.3)

4. Understand the first law of thermodynamics, and terms related to this concept, including endothermic and exothermic reactions, energy change, and enthalpy change. (Sec. 6.4)

5. If given an set of equations with associated enthalpy change values, use Hess's Law to calculate the enthalpy change for a reaction that is a combination of those given. (Secs. 6.5 & 6.6)

6. Know the special properties of state functions and recognize which of the thermodynamic variables discussed so far are state functions. (Sec. 6.7)

7. Know the standard conditions for thermochemical calculations and understand what is meant by the standard enthalpy of formation or reaction. This includes knowing that the standard enthalpy of an element in its standard state must be zero.

   Be able to use an appropriate table of standard enthalpies of formation with the equation

$$\Delta H^{o}_{rxn} = \Sigma \ \Delta H^{o}_{f} \ (\text{products}) - \Sigma \ \Delta H^{o}_{f} \ (\text{reactants})$$

to calculate the standard enthalpy change for reactions. (Sec. 6.8)

8. Understand the basic experimental procedures of calorimetry, that is, the method used to measure heats of reactions in the laboratory. (Sec. 6.9)

108

9. Be aware of the energy sources which are currently used most widely in this country and some possible alternative energy sources. (Sec. 6.10)

ARMCHAIR EXERCISES
(More hints are at the end of the chapter.)

1. Stretch out a heavy rubber band and hold it in place for several seconds to allow it to come to a constant temperature. Now, keeping the band stretched, place it against your upper lip and feel the temperature. (Your upper lip is quite sensitive to temperature changes.) Allow the band to contract, and immediately place it against your lip once again. Do you feel a temperature change? Is this a spontaneous or nonspontaneous process? Is it exothermic or endothermic? Is this an unusual combination?

2. Why is it that you would be badly burned if you put your hand into a pot of boiling water, but you don't even feel much discomfort if you reach into the oven to retrieve a dish, even though the temperature is probably almost twice as high in the oven as it is in the boiling water?

3. If you place water into a paper cup and then heat it with a match, the cup will not burn, even though it may smolder. When you do the same thing with a styrofoam cup, the heat melts the cup. Why does this difference occur? (IF YOU ACTUALLY ATTEMPT THIS DEMONSTRATION, BE SURE TO DO IT OVER A SINK AND BE CAREFUL NOT TO BURN YOURSELF.) HINT: What difference do you observe between the two cups when there is a hot liquid inside the cup?

CONCEPT TEST
(Answers are provided at the end of the chapter.)
1. The science of heat or energy flow in chemical reactions is called _____.

2. _____ is defined as the capacity to do work.

3. The amount of heat required to raise the temperature of 1.00 gram of pure liquid water from

14.5°C to 15.5°C is called a(n) _____.

4. _____ is the name of the SI unit of energy. One calorie is equivalent to _____ of these units.

5. The ratio of the heat supplied to 1.00 gram of a substance to the consequent rise in the substance's temperature is called the _____.

6. Changes of state always occur at constant _____.

7. The quantity of heat required to melt a substance at its melting point is called the _____.

8. When heat must be transferred into a system to maintain constant temperature as the process occurs, the process is said to be _____; when heat is transferred from the system to the surroundings, the process is _____.

9. The first law of thermodynamics states that the total amount of energy in the universe is _____.

10. The heat transferred into (or out of) a system at constant pressure is called the _____.

11. When heat is transferred at constant pressure to the surroundings by an exothermic reaction, what is the sign for ΔH? _____

12. The value of a(n) _____ depends only on the state of a system; therefore, the change in value of such a function is determined only by its initial and final value.

13. The standard state of an element or compound is measured at a pressure of _____ and a

temperature of _____.

14. Standard enthalpies of formation for the elements in their standard states have a value of _____.

15. Around the middle of this century, coal was replaced by _____ as the main fuel for the industrialized nations.

STUDY HINTS

1. The properties of state functions are the key to most of the problem-solving methods described in this chapter; therefore, it's critical that you understand the meaning of state function and be able to recognize the common state functions.

2. Thermodynamic symbols often include subscripts and superscripts that represent important information about the conditions, such as temperature and pressure. For example, the superscript o on the symbol ΔH^{o} should tell you the temperature, the pressure, and the physical state of the substance involved. Don't overlook this valuable source of information.

3. Some students have difficultly distinguishing the different problem types in this chapter. It may be helpful to note that learning goal 3 usually refers to systems undergoing physical change, that is, melting, freezing, or temperature change. Learning goals 5 and 7 describe closely related problems, but in the first case the data is usually a list of equations with the corresponding enthalpy values and in the second case the data is provided as a table of standard enthalpy values. Try to identify these three different situations as you look at the problems in the text and in this guide.

4. Be sure that all of the equations are balanced before you begin to do a thermochemistry problem. Failure to do this can waste time (if you notice the error later on) or cause your solution to be incorrect.

PRACTICE PROBLEMS

1. **(L1)** The box of a popular breakfast cereal states that one serving provides 50 Calories (without milk). How much energy is this if measured in joules? (Hint; Notice that there is a capital C on the energy unit, so it's really a kilocalorie!)

2. **(L2 and L3)** How many joules of energy are required to change 100.0 grams of ice at $0.0°C$ to liquid water at $40.0°C$. The heat of fusion for water is 333 J/g.

3. **(L7)** Calculate the Standard Enthalpy of Combustion for the reaction

$$C_2H_4(g) \ + \ O_2(g) \ \rightarrow \ CO_2(g) \ + \ H_2O(g)$$

Based on the following Standard Enthalpies of Formation: $C_2H_4(g)$ +52.3 kJ/mol; $CO_2(g)$ -393.5 kJ/mol; and $H_2O(g)$ -241.8 kJ/mol

112

4. (L5) Cyanamide, CH_2N_2, is a weak acid that is sometimes used as a fertilizer. Calculate the standard enthalpy of formation for cyanamide, given the following standard enthalpies of reaction:

$$CH_2N_2(s) + 3/2\ O_2(g) \rightarrow CO_2(g) + H_2O(\ell) + N_2(g) \quad \Delta H_1 = -741.4\ \text{kJ/mol}$$

$$C(s) + O_2(g) \rightarrow CO_2(g) \quad \Delta H_2 = -393.5$$

$$H_2(g) + 1/2\ O_2(g) \rightarrow H_2O(\ell) \quad \Delta H_3 = -285.8$$

5. (L7) Phosgene, $COCl_2(g)$, and hydrogen sulfide are both very poisonous, but when they are heated together the product is carbon disulfide, an evil-smelling gas that is much less toxic than either of the reactants. Calculate the standard enthalpy of formation for phosgene given that the standard enthalpy of reaction is -49.0 kJ for

$$COCl_2(g) + 2\ H_2S(g) \rightarrow 2\ HCl(g) + H_2O(g) + CS_2(g)$$

and using the following standard enthalpies of formation:

	ΔH_f^o		ΔH_f^o
$CS_2(g)$	117.4 kJ/mol	$H_2S(g)$	-20.63 kJ/mol
$HCl(g)$	- 92.3	$H_2O(g)$	-241.8

6. **(L8)** If 1.15 grams of glucose (molar mass = 180.2 g/mol) is burned in a calorimeter, the temperature increases from 23.40°C to 27.21°C. If the heat capacity of the calorimeter is 958 j/°C and the calorimeter contains 899.5 grams of water, calculate the amount of heat given off per mole of glucose under these conditions.

7. **(L5)** Coke reacts with steam to produce a mixture called coal gas, which can be used as a fuel or as a starting material for other reactions. The equation for the production of coal gas is

$$2 \ C(s) \ + \ 2 \ H_2O(g) \rightarrow CH_4(g) \ + \ CO_2(g)$$

Determine the Standard Enthalpy Change for this reaction based on the following Standard Enthalpies of Reaction:

$C(s) \quad + \quad H_2O(g)$
$\rightarrow \quad CO(g) \quad + \quad H_2(g) \qquad \Delta H^\circ = +131.3 \ kJ$

$CO(g) \quad + \quad H_2O(g)$
$\rightarrow \quad CO_2(g) \quad + \quad H_2(g) \qquad \Delta H^\circ = - \ 41.2$

$CH_4(g) \quad + \quad H_2O(g)$
$\rightarrow \quad 3 \ H_2(g) \ + \quad CO(g) \qquad \Delta H^\circ = +206.1$

PRACTICE PROBLEM SOLUTIONS

1. Remember that the calories used in this case are one thousand times as large as the calorie units we normally use and one of these "small" calories is 4.184 joules. Thus the conversion is

$$\text{joules} = 50 \text{ Cal} \times 1000 \text{ cal/Cal} \times 4.184 \text{ J/cal}$$
$$\underline{\text{joules} = 2 \times 10^5 \text{ joules}}$$

2. First, calculate how much heat is needed to melt the ice at the normal melting point, $0°C$.

$$q = \Delta H_{fus} \times m = 333 \text{ J/g} \times 100.0 \text{ g} = 33300 \text{ J}$$

Next, determine how much heat is required to heat the liquid water 40 degrees Celsius.

$$q = c \times m \times \Delta t$$
$$= 4.184 \text{ J/g·}°C \times 100.0 \text{ g} \times 40.0°C$$
$$q = 16736 \text{ J}$$

Finally, add heat needed for the two processes and round to the correct number of significant figures.

$$q = 33300 \text{ J} + 16736 \text{ J}$$
$$\underline{q = 50.0 \text{ kJ}}$$

3. First, balance the equation

$$C_2H_4(g) + 3 O_2(g) \rightarrow 2 CO_2(g) + 2 H_2O(g)$$

Now combine the standard enthalpy of formation values provided:

$$\Delta H°_{rxn} = \Sigma \Delta H°_f \text{ (products)} - \Sigma \Delta H°_f \text{ (reactants)}$$

$$\Delta H° = 2 \Delta H°_f [CO_2(g)] + 2 \Delta H°_f [H_2O(g)] - \Delta H°_f [C_2H_4(g)]$$

$$\Delta H° = 2(-393.5 \text{ kJ/mol})$$
$$+ 2(-241.8 \text{ kJ/mol}) - (52.3 \text{ kJ/mol})$$

$$\underline{\Delta H° = -1320 \text{ kJ}}$$

4. The equation for the formation of cyanamide from the elements in their standard states is

$$C(s) + H_2(g) + N_2(g) \rightarrow CH_2N_2(s)$$

To obtain this equation, reverse the first equation

$$CO_2(g) + H_2O(\ell) + N_2(g) \rightarrow CH_2N_2(s) + 3/2\ O_2(g)$$

Use the second equation as written

$$C(s) + O_2(g) \rightarrow CO_2(g)$$

And use the third equation as written

$$H_2(g) + 1/2\ O_2(g) \rightarrow H_2O(\ell)$$

Adding these three equations produces the desired reaction, thus the enthalpy change for this reaction must be equal to

$$\Delta H = -\Delta H_1 + \Delta H_2 + \Delta H_3$$

$$\Delta H = -(-741.4\ kJ) + (-393.5\ kJ) + (-285.8\ kJ)$$

$$\underline{\Delta H = +62.1\ kJ/mol}$$

5. The equation is already balanced, so substitute into the Hess equation:

$$\Delta H^{\circ}_{rxn} = \Sigma\ \Delta H^{\circ}_f\ (products) - \Sigma\ \Delta H^{\circ}_f\ (reactants)$$

$$\Delta H^{\circ} = 2\ \Delta H^{\circ}_f\ [HCl(g)] + 2\ \Delta H^{\circ}_f\ [H_2O(g)] - \Delta H^{\circ}_f\ [CS_2(g)]$$
$$+ 2\ \Delta H^{\circ}_f\ [COCl_2(g)] - 2\ \Delta H^{\circ}_f\ [H_2S(g)]$$

$$-49.0\ kJ = 2(-92.3 kJ/mol) + (-241.8\ kJ/mol)$$
$$+ (117.4\ kJ/mol) - \Delta H^{\circ}_f\ [COCl_2(g)]$$
$$- 2(-20.63\ kJ/mol)$$

$$\Delta H^{\circ}_f\ [COCl_2(g)] = \underline{-218.7\ kJ/mol}$$

6. The heat absorbed by the bomb is given by

$$q_{bomb} = (C_{bomb})(\Delta t) = 958 \text{ J/}^\circ\text{C} \times (27.21-23.40)$$

$$= 3.65 \times 10^3 \text{ J}$$

The heat absorbed by the water equals

$$q_{water} = (4.184 \text{ J/g} \cdot ^\circ\text{C})(m_{water})(\Delta t)$$

$$= (4.184 \text{ J/g} \cdot ^\circ\text{C})(899.5 \text{ g})(27.32-23.40)$$

$$= 14.3 \times 10^3 \text{ J}$$

To find the heat generated by the combustion, add these two heat values and reverse the sign.

Total heat by 1.15 g of glucose $= -18.0 \times 10^3$ J

$$= -18.0 \text{ kJ}$$

Now find the heat evolved per gram of glucose and multiply by the molar mass of this compound.

$$\Delta H = \frac{(-18.0 \text{ kJ})(180.2 \text{ g/mol})}{1.15 \text{ g}}$$

$$\underline{\Delta H = -2820 \text{ kJ/mol}}$$

7. To obtain the equation specified, double the first reaction

$$2 \times [C(s) + H_2O(g) \rightarrow CO(g) + H_2(g)]$$

use the second reaction as written

$$CO(g) + H_2O(g) \rightarrow CO_2(g) + H_2(g)$$

and reverse the third reaction.

$$3 H_2(g) + CO(g) \rightarrow CH_4(g) + H_2O(g)$$

Adding these three reactions and canceling will produce

$$2 C(s) + 2 H_2O(g) \rightarrow CH_4(g) + CO_2(g)$$

• Negative formal charges should reside on the most electronegative atoms.

• Like formal charges on adjacent atoms should be avoided, because structures having like formal charges on adjacent atoms are not likely to be stable.

In using the guideline to minimize formal charges, it is common to draw Lewis structures with multiple bonds that place more than four pairs of electrons about atoms from the third period or below, atoms such as S and P. This is acceptable, because these atoms have access to vacant d orbitals having the same principal quantum number as their s and p valence orbitals and are known to form molecules and ions in which they have more than an octet in their valence shell.

Atom Partial Charges

Oxidation numbers are based on assigning shared electrons to the more electronegative bonded atoms whereas formal charges are based on assigning shared electrons equally to bonded atoms. Each approach provides numbers suitable for certain uses, but neither approach gives the best estimates of charges on atoms in molecules and polyatomic ions, because bonded electrons are usually shared unequally in such species.

Professor Leland Allen of Princeton University has developed a more reasonable means of estimating atom charges that allows for unequal sharing of bonded electrons. The atom charges that are obtained using Allen's method are fractional charges and are therefore called **partial charges.** *According to Allen,*

partial charge = group number of A - number of lone pair
  on atom A   electrons on A - $(\chi_A/\Sigma\chi)$(number of bonding
              electrons shared by A)   (9c)

where $\Sigma\chi$ is the sum of the electronegativities of atom A and the atom to which it is bonded.

• In the case of LiF, the partial charge of Li = 1 - 0 - $[\chi_{Li}/(\chi_{Li} + \chi_F)](2) = 1 - 0 - (1.0/5.0)(2) = 0.60+$.

• In the case of $Cl_2C=O$, the partial charge of C = 4 - 0 - $[\chi_C/(\chi_C + \chi_{Cl})](2) - [\chi_C/(\chi_C + \chi_{Cl})](2) - [\chi_C/(\chi_C + \chi_O)](4) = 4 - 0 - (2.5/5.5)(2) - (2.5/5.5)(2) - (2.5/6.0)(4) = 0.52+$.

In both of these cases, the sum of the partial charges for all the atoms must equal zero, because there is no overall charge

---

5. The combustion of acetylene, C_2H_2, is used to produce high temperatures for welding torches. Calculate the Standard Enthalpy of Reaction for the balanced equation

$$2\ C_2H_2(g)\ +\ 5\ O_2(g)\ \rightarrow\ 4\ CO_2(g)\ +\ 2\ H_2O(\ell)$$

Given the Standard Enthalpies of Reaction for the following:

$$2\ C(graphite) +\ H_2(g)\ \rightarrow\ C_2H_2(g)\qquad \Delta H_1\ =\ 227\ kJ$$

$$C(graphite)\ +\ O_2(g)\ \rightarrow\ CO_2(g)\qquad \Delta H_2\ =\ -394$$

$$H_2(g)\qquad +\ 1/2\ O_2(g)\ \rightarrow\ H_2O(\ell)\qquad \Delta H_3\ =\ -286$$

6. Write the equation for the formation of $H_2O_2(g)$, gaseous hydrogen peroxide, from the elements at $25°C$ in their standard states, and then calculate the standard enthalpy of formation for this compound based on the following reactions:

$$2\ H(g)\ +\ 2\ O(g)\ \rightarrow\ H_2O_2(g)\qquad \Delta H°\ =\ -1071\ kJ$$

$$O_2(g)\ \rightarrow\ 2\ O(g)\qquad \Delta H°\ =\ 498.3$$

$$2\ H(g)\ +\ O(g)\ \rightarrow\ H_2O(g)\qquad \Delta H°\ =\ -\ 926.9$$

$$H_2(g)\ +\ 1/2\ O_2(g)\ \rightarrow\ H_2O(g)\qquad \Delta H°\ =\ -\ 241.8$$

7. When 0.500 grams of sucrose (molar mass = 342.3g/mol) is burned in a calorimeter, the temperature increases from $21.11°C$ to $23.42°C$. Calculate the heat released per mole of sucrose, if the heat capacity of the calorimeter is 0.852 kJ/$°C$ and the calorimeter contains 650.0 grams of water.

8. Gasoline might be conserved by marketing a blend of ethanol and gasoline as an auto fuel. Write the equation for the combustion of ethanol, $C_2H_5OH(g)$, and use these standard enthalpies of formation to calculate the standard enthalpy of combustion.

$C_2H_5OH(g)$ -235.1 kJ/mol $CO_2(g)$ -393.5 kJ/mol

$H_2O(\ell)$ -285.8

9. The organic compound ethylene, C_2H_4, reacts with hydrogen to form ethane, C_2H_6:

$$H_2C=CH_2(g) \; + \; H_2(g) \; \rightarrow \; H_3C-CH_3(g)$$

Calculate the enthalpy of reaction for this reaction based on the following:

$H_2C=CH_2(g) + 3 \; O_2(g)$
$$\rightarrow 2 \; CO_2(g) + 2 \; H_2O(g) \quad \Delta H^\circ = -1323 \text{ kJ}$$

$2 \; H_2(g) \quad + \; O_2(g) \; \rightarrow \; 2 \; H_2O(g) \quad \Delta H^\circ = -483.6$

$H_3C-CH_3(g) + 7/2 \; O_2(g)$
$$\rightarrow 2 \; CO_2(g) + 3 \; H_2O(g) \quad \Delta H^\circ = -1428$$

COMMENTS ON ARMCHAIR EXERCISES

1. The process is spontaneous, but it's also endothermic. This is an unusual combination, which will be discussed later in the course.

2. Think about the number of molecules you will encounter in these two cases. Does that suggest a reason why more heat will be transferred to your finger from the hot water?

3. Water has a high heat capacity, and the heat passes through the paper cup readily. The water inside absorbs heat, which would otherwise burn the cup. Why is the polystyrene different? Think about how much heat passes through the cup when there is a not beverage inside. Does this give you a clue?

CONCEPT TEST ANSWERS

1. thermochemistry   2. energy
3. calorie   4. joule, 4.184 joule/cal
5. specific heat capacity
6. temperature   7. heat of fusion
8. endothermic, exothermic
9. constant   10. enthalpy change
11. negative   12. state function
13. 1 atm, $25^\circ C$ 14. zero
15. petroleum and natural gas

PRACTICE TEST ANSWERS

1. (L7) a. C(graphite)   b. $F_2(g)$

   c. $CO_2(g)$ d. $H_2O(\ell)$

2. (L7) −1127 kilojoules

3. (L5) $N_2(g) + 2 H_2(g) \rightarrow N_2H_4(g)$ +95 kJ

4. (L2 and L3) 10.9 kJ

5. (L5) −2602 kJ

6. (L5) $H_2(g) + O_2(g) \rightarrow H_2O_2(g)$ −112.4 kJ

7. (L8) −5650 kJ/mol

8. (L7) $C_2H_5OH(g) + 3 O_2(g) \rightarrow 2 CO_2(g) + 3 H_2O(g)$

   −1409.3 kJ/mol

9. (L5) −137 kJ

CHAPTER 7
ATOMIC STRUCTURE

CHAPTER OVERVIEW

As you may have noticed, a special feature of this chapter is the historical method of presentation. There is a reason for this. Often the experiments or equations developed by a given scientist are closely associated with his name, for example, the de Broglie equation or the Schrödinger equation. Therefore, it's important for you to remember the nature of the work done by each individual. This should also help you to visualize the complex set of developments in several different fields of science that all combined to produce the theory of atomic structure.

Quantization is one of the key ideas of modern science, but it is quite contrary to what you observe in everyday life. In fact, there was a great deal of resistance to the idea when it was first proposed by Einstein and Planck. Perhaps the closest analogy is currency. Items are rarely priced to the nearest half or quarter of a cent because there is no coin of that size. Similarly, energy is absorbed and emitted only in some multiple of the basic energy amount, the quantum.

Another theme that runs through this chapter is the development of quantum numbers. Originally created by Bohr in order to explain why certain atomic orbits didn't radiate energy, the concept was expanded by Schroedinger, and has since been applied to many different energy measurements. Learning the rules is essential, since only certain quantum number values are allowed.

Schroedinger's quantum numbers describe the energy, shape, and orientation of the atomic orbitals. The concept of an orbital can be rather confusing, since an orbital is really just a mathematical description of where a pair of electrons might be found. An orbital has no real existence unless it contains electrons and, at best, represents only an approximation of electron positions, but the concept is very useful.

122

ARMCHAIR EXERCISE

1. Since diffraction patterns often consist of a series of dots, students may think that the dots are actually images of the particles that created the diffraction pattern. Can you explain why this response is correct or incorrect?

It is easy to make a simple diffraction pattern. Make a slit by holding two of your fingers together, then look at a light. If you adjust the width of the gap between your fingers, you will notice a series of alternating light and dark lines. As you increase the gap between your fingers these lines will eventually disappear. The light and dark lines are a diffraction pattern.

2. What color are electrons? You have probably noticed that neither your instructor or the textbooks ever mentions the color of electrons. Can you explain this?

3. Do atomic orbitals really have edges? Many of the pictures of orbitals show them to have very distinct edges. Is this accurate? Look carefully at the illustrations is section 7.6 of the text as you think about this question.

LEARNING GOALS

1. Much of our information about atomic structure has been gained from observing the way in which matter absorbs or emits electromagnetic radiation. Therefore, to understand this chapter you must learn the fundamental wave properties of electromagnetic radiation, including making conversions with the basic equation
$$\lambda \upsilon = c$$
(Sec. 7.1)

2. In addition to the traditional wave picture of electromagnetic radiation, scientists such as Planck and Einstein proposed that electromagnetic radiation had a particle nature, as defined by the equation
$$E = h\upsilon$$
You should also understand the basic ideas of the

Quantum Theory, since it's also a crucial component of the modern theory of atomic structure. (Sec. 7.2)

3. The study of atomic line spectra is a third major component in the development of the theory of atomic structure. You should be able to use the Rydberg equation and understand how it served as a basis for Bohr's picture of the atom.

$$\Delta E = - (Rhc/n_f^2) - (Rhc/n_i^2)$$

You should also be able to calculate the energies of atomic energy states and spectral lines for hydrogen gas. (Sec. 7.3)

4. The next step in the development of the quantum theory was provided by de Broglie, who proposed that there was a specific wavelength associated with a moving electron. Understand how de Broglie's ideas served as a basis for Heisenberg's development of the uncertainty principle and ultimately served as the basis for quantum mechanics. You should also be able to use de Broglie's equation

$$\lambda = h/mv$$

(Sec. 7.4)

5. According to the ideas of quantum mechanics, quantum numbers play a key role in describing the energy of the electrons in an atom. It is essential that you thoroughly learn the rules that determine which combinations of quantum number values are permitted as well as the letter designations, s, p, d, f, etc. for the ℓ values. In the next chapter, you will see how these ideas provide a basis for the understanding of atomic structure. (Sec. 7.5)

6. As has been seen in the previous section, there are several different types of atomic orbitals and several different ways to represent the shapes of these orbitals. You should be able to recognize these various representations for at least s, p, and d orbitals. You should also understand what a node is and how to determine the number of nodes for a given type of orbital. (Sec. 7.6)

CONCEPT TEST

(Answers are provided at the end of the chapter.)

1. The distance between comparable points (i.e. peak to peak or trough to trough) of a wave is called the _____.

2. The number of waves passing a stationary point in a specified time interval is called the

_____.

3. The maximum height of a wave, as measured from the axis of propagation, is the _____.

4. The product of the wavelength multiplied times the _____ for any wave phenomena is equal to the velocity of propagation for the waves.

5. In a vacuum, the velocity of propagation of all electromagnetic radiation is _____.

6. If you tie down a string at both ends and pluck it (like a guitar or violin), the type of wave motion or vibration that result is called a(n)

_____.

7. Plank suggested that all energy gained or lost by an atom must be some integer multiple of a minimum amount of energy called a(n) _____.

8. The photoelectric effect occurs when light strikes the surface of a metal and _____ are ejected.

9. Based on studies of the photoelectric effect, Einstein proposed that light could be described as having particle-like properties, due to the presence of massless particles that are now called

_____.

10. A spectrum that contains light of all colors is

called a(n) _____ spectrum.

11. Emission line spectra are produced by materials in which of the three states of matter, gas, liquid, or solid? _____

12. An atom with its electrons in the lowest possible energy levels is said to be in the _____.

13. According to Bohr, atomic line spectra occur when electrons move between _____.

14. De Broglie suggested that any moving particle has an associated wavelength, but that for large particles, like baseballs, the size of the wavelength is _____.

15. According to Heisenberg's uncertainty principle, if one attempts to simultaneously measure the energy and position of an electron in an atom, the more exactly the energy is measured, the greater will be the _____ in the position measurement.

16. The electron density is the _____ of finding an electron within a given region of space.

17. The matter waves for the allowed energy states of an atom are called _____.

18. In order to solve the Schrödinger equation for an electron in three dimensions, it is mathematically necessary to introduce three numbers having integer values called _____.

19. When two or more electrons have the same value of the quantum number, n, they are said to be in the same _____.

126

20. Each quantum number has a different meaning. **The principal quantum number, n,** is a measure of the most probable _____ from the nucleus; ℓ, **the angular momentum quantum number,** is related to the _____ of the electron orbitals, and m_ℓ, **the magnetic quantum number,** specifies in which _____ within a subshell the electron is located.

21. a. When **n** = 5, which of the following values are possible for ℓ, ℓ = -1, 0, 3, 5, or 6? _____

b. When ℓ = 3, which of the following values are possible for m_ℓ, m_ℓ = -2, 0, 3, or 4. _____

22. a. When ℓ = 3, the maximum number of orbitals of this type in a given electron shell is _____ (how many?), and a letter designation of _____ (choose from s,p,d, or f) is assigned.

b. When ℓ = 1, the maximum number of orbitals of this type in a given electron shell is _____,and a letter designation of _____ is assigned.

23. The size of an s orbital increases as the value of the quantum number _____ increases.

24. A region of a probability density graph where the probability of finding the electron is zero is called a(n) _____.

25. One nanometer is _____ meters.

STUDY HINTS

1. Students are sometimes confused by the reciprocal time units used for frequencies, but these units do make sense. By comparison, if someone asks you how often you eat regular meals, you may answer, "three times per day." If we write

this statement in the same way as frequency results, it would become 3 day^{-1}. When we say that the frequency of a wave phenomena is 1,000,000 hertz, we are only saying that 1,000,000 waves per second would go past a stationary observation point.

2. Having read the chapter, you may well wonder whether an electron really is a wave or a particle. One way to satisfy yourself is to consider that we are not really arguing about the nature of an electron, but rather about which of two possible explanations works best. Our problem is apparently that neither explanation works all of the time, that is, the true nature of the electron is not exactly like either of our theories.

3. There is no substitute for carefully memorizing the rules that determine the possible quantum number values. It is impossible to understand atomic structure unless you have thoroughly learned these rules. In addition, it is essential to memorize the correspondence between the numerical values of ℓ and the letter values, i.e. s, p, d, f, etc. Until you know these relationships well, it will be difficult to follow any description of what the quantum numbers mean.

PRACTICE PROBLEMS

1. (L1) a. For each pair of types of electromagnetic radiation given, circle the type having the higher energy.

    ultraviolet or infrared

    x-rays or γ-rays

    blue light or yellow light

b. For each pair of types of electromagnetic radiation given circle the type having the greater wavelength.

    microwave or infrared

    AM radio or FM radio

    x-rays or ultraviolet

2. **(L2)** According to the Bohr Theory, the seventh line of the Paschen series in the emission spectrum of hydrogen gas occurs when an electron goes <u>from</u> an orbit having an n value of _____ to an orbit having an n value of _____ .

3. **(L5)** a. When $n = 5$, what are the possible values for ℓ? _____

b. When $\ell = 5$, what are the possible values for m_ℓ? _____

c. When $n = 2$, what are the possible values for m_ℓ? _____

4. **(L5)** State whether each of the following combinations of quantum numbers is allowed or not according to the rules you learned in the text? If it is not allowed, explain why.

a. $n = 3$, $\ell = 2$, and $m_\ell = -2$ _____
b. $n = 6$, $\ell = 0$, and $m_\ell = 1$ _____
c. $n = 2$, $\ell = 1$, and $m_\ell = 0$ _____
d. $n = -1$, $\ell = 1$, and $m_\ell = 1$ _____

5. **(L5)** a. What is the maximum number of p orbitals that are found in a given electron shell of an atom? _____ b. What is the maximum number of d orbitals that are found in a given electron shell of an atom? _____

6. **(L4)** According to Heisenberg's uncertainty principle, we cannot simultaneously determine the position and momentum of an object. Why don't we observe this effect in our everyday life?

7. **(L1 and L2)** Calculate (a) the frequency and (b) the energy of a photon of light having a wavelength of 590 nanometers.

8. **(L4)** Calculate the de Broglie wavelength (in nanometers) of a golf ball moving with a speed of 145 km/s, if the mass of the golf ball is 45.7 grams.

9. **(L3)** Calculate the energy and the frequency of the fourth line in the Paschen Series of the emission spectrum of hydrogen atom in kJ/mole. The electrons that produce this series of lines move from higher energy states to the n = 3 state.
$(R = 1.0974 \times 10^7 \text{ m}^{-1})$

PRACTICE PROBLEM SOLUTIONS

1. a. The radiation having the higher energy is

    ultraviolet

    γ-rays

    blue light

   b. The radiation having the larger wavelength is

    microwave

    AM radio

    ultraviolet

2. from n = 10 to n = 3

3. a. +4, +3, +2, +1, and 0
   b. +5, +4, +3, +2, +1, 0, -1, -2, -3, -4, and -5
   c. $\ell = 1$, $m_{\ell} = +1$, 0, -1
      $\ell = 0$, $m_{\ell} = 0$
4. b. is not allowed
      because m_{ℓ} cannot be greater than ℓ
   d. is not allowed
      because n cannot have negative values
5. a. three p orbitals, b. five d orbitals
6. The uncertainty for objects that we encounter in everyday life is much smaller than our normal ability to detect. The uncertainty is still there; we just aren't aware of it.
7. a. Convert the wavelength to meters and substitute into the equation that relates wavelength and frequency.

$$\upsilon = \frac{c}{\lambda} = \frac{(3.00 \times 10^8 \text{ m/s})}{(590 \text{ nm})(1 \times 10^{-9} \text{ m/nm})}$$

$$\underline{\upsilon = 5.08 \times 10^{14} \text{ s}^{-1}}$$

b. Using this frequency in the Planck relationship

$$E = h\upsilon = (6.63 \times 10^{-34} \text{ J} \cdot \text{s})(5.08 \times 10^{14} \text{ s}^{-1})$$

$$\underline{E = 3.37 \times 10^{-19} \text{ J}}$$

8. Convert the mass to kilograms and the speed to meters/second, then use the de Broglie wavelength equation

$$\text{wavelength of ball} = \lambda = \frac{h}{mv}$$

$$\lambda = \frac{(6.6262 \times 10^{-34} \text{ J} \cdot \text{s})(1000 \text{ g/kg})}{(45.7 \text{ g})(145 \text{ km/s})(1000 \text{ m/km})}$$

$$\lambda = 1.00 \times 10^{-37} \text{ m}$$

Convert to nanometers

$$\lambda = (1.00 \times 10^{-37} \text{ m})(1 \times 10^9 \text{ nm/m})$$

$$\underline{\lambda = 1.00 \times 10^{-28} \text{ nm}}$$

9. The energy difference is given by

$$\Delta E = - (Rhc/n_2^2) - (Rhc/n_1^2)$$

$$\Delta E = - (Rhc/7^2) - (Rhc/3^2)$$

$$\Delta E = -(1.0974 \times 10^7 m^{-1})(6.6261 \times 10^{-34} J \cdot s)$$
$$\times (2.9979 \times 10^8 m/s)(-0.0907)$$

$$\Delta E = 1.9772 \times 10^{-19} \ J/atom$$

Multiply by Avogadro's number and convert to kilojoules

$$\Delta E = (1.9772 \times 10^{-19} \ J/atom)$$
$$\times (6.022 \times 10^{23} \ atoms/mol)(1kJ/1000J)$$

$$\underline{\Delta E = 119.07 \ kJ/mole}$$

Now convert this energy into a frequency

$$\upsilon = \frac{\Delta E}{h} = \frac{(-119.07 \ kJ/mol)(10^3 J/kJ)}{(6.022 \times 10^{23} \ electrons/mol)(6.626 \times 10^{-34} \ J \cdot s)}$$

$$\underline{\upsilon = 2.984 \times 10^{14} \ s^{-1}}$$

PRACTICE TEST (40 min.)

1. a. What is the maximum number of p orbitals that are found in a given electron shell of an atom? b. What is the maximum number of g orbitals?

2. Which of the following combinations of quantum numbers is not allowed according to the rules discussed in the textbook.

a. $n = 0$, $\ell = 2$, and $m_\ell = 1$.
b. $n = 4$, $\ell = 0$, and $m_\ell = 0$.
c. $n = 3$, $\ell = 2$, and $m_\ell = 3$.
d. $n = 5$, $\ell = -4$, and $m_\ell = 0$.

3. Listed on the left below are the names of some of the scientists who made major contributions to

the theory of atomic structure. On the right is a list of those contributions. Match the idea with the scientist by placing the letter from the right hand column in the appropriate space on the left.

_____ Balmer

_____ Bohr

_____ de Broglie

_____ Einstein

_____ Heisenberg

_____ Thomson

_____ Planck

a. position and momentum cannot be measured exactly and simultaneously

b. The wave properties of the matter

c. explanation of atomic spectra

d. mass to charge ratio for the electron

e. explained the photoelectric effect

f. quantization of energy emission and absorption

g. mathematical relation that predicts the first three lines of the visible emission spectrum for hydrogen gas

4. According to the Bohr Theory, the fifth line of the Lyman series in the emission spectrum of hydrogen gas occurs when an electron goes _from_ an n value of _____ to an n value of _____.

5. If the wavelength is 7.00×10^{-5} cm for a certain photon of light, what is the energy of this photon in joules?

6. Infrared radiation has a wavelength of 1.2×10^3 nanometers. a. What is the frequency of this radiation? b. What is the energy in joules of one photon of this radiation? c. What is the energy in joules of 1.00 moles of photons of this radiation?

7. If an electron has a mass of 9.109×10^{-31} kilograms and is moving with a velocity 10.0% of the velocity of light, what is its de Broglie wavelength?

8. Match the terms in the left hand column with the most appropriate description from the right hand column.

_____ photoelectric effect

_____ nodal plane

_____ stationary states

_____ wave function

_____ uncertainty principle

a. only certain quantized atomic energy levels are allowed

b. electrons are ejected when light strikes the surface of some metals

c. the more exactly the momentum of a particle is known, the less exactly the position can be measured

d. region where the probability of finding an electron is zero

e. solutions of the Schrödinger equation

9. The Pfund Series of atomic spectral lines for hydrogen gas are the result of electronic transitions that terminate in the energy level where n = 5. What is the energy of the fourth line of the Pfund Series in J/atom? in kJ/mole?

10. According to the quantum number rules discussed in this chapter, which of the following orbital types would not be possible: 1p, 4s, 4f, 2d, 3g?

11. How many planar nodes are associated with each of the following orbitals? (a) 3d, (b) 4p, and (c) 5s.

COMMENTS ON ARMCHAIR EXERCISES

1. Notice that when you look at the diffraction pattern between your fingers, the image duplicates the slit that you have formed. Are you really seeing a picture of your fingers, or is the image

simply formed by the space between the fingers. Does this suggest an answer to the question of whether or not you are really seeing atoms?

2. In order for something to display a color, it must interact with light rays. Does this seem likely for electrons? If you are in doubt, reread the section on the Heisenberg Uncertainty Principle.

3. The "edges" that we show for orbitals are really just a distance at which the probability of observing an electron is less than some arbitrary value, so they aren't really as definite as you see in pictures. Does this suggest some interesting ideas about atoms?

CONCEPT TEST ANSWERS

1. wavelength
2. frequency
3. amplitude
4. frequency
5. 2.998×10^8 m/s
6. standing or stationary wave
7. quantum
8. electrons
9. photons
10. continuous
11. gas
12. ground state
13. quantized energy states or stationary states
14. too small to measure
15. uncertainty
16. probability
17. orbitals
18. quantum numbers
19. electron shell
20. orbital size or diameter, shape, orbital
21. a. 0 and 3 for ℓ,
    b. -2, 0, and 3 for m_ℓ.
22. a. seven f orbitals
    b. three p orbitals
23. n
24. node
25. 1×10^{-9} meters

PRACTICE TEST ANSWERS

1. (L5) a. $\ell = 1$, so three orbitals are possible.
   b. $\ell = 4$, so nine orbitals are possible.
2. (L5) The combinations not allowed are
   a. (n cannot equal zero),
   c. (m_ℓ cannot be greater than ℓ.), and
   d. (ℓ cannot have a negative value.)

3. Balmer - g, Bohr - c, de Broglie - b, Einstein - e, Heisenberg - a, Thomson - d, and Planck - f.

4. (L3) <u>from</u> n = 6 <u>to</u> n = 1

5. (L2) 2.84×10^{-19} J

6. (L1 and L2) a. 2.5×10^{14} s^{-1}

   b. 1.7×10^{-19} joules   c. 1.0×10^{5} joules

7. (L4) 2.426×10^{-11} meters

8. (L2, L4, and L5)

photoelectric effect -   b. electrons are ejected when light strikes a metal surface

nodal plane -   d. region where the probability of finding an electron is zero

stationary states -   a. only certain quantized atomic energy levels are allowed

wave function -   e. solutions of the Schrödinger equation

uncertainty principle -   c. the more exactly the momentum of a particle is known, the less exactly the position can be measured

9. (L3) 6.028×10^{-20} J/atom, 36.30 kJ/mol

10. (L6) 1p, 2d, and 3g are impossible.

11. (L6) (a) two, (b) one, (c) none.

CHAPTER 8
ATOMIC ELECTRON CONFIGURATIONS
AND CHEMICAL PERIODICITY

CHAPTER OVERVIEW

It can be argued that one of the greatest scientific discoveries of this century has been the description of the electronic configurations of the elements and the realization that the chemical behavior of the elements is a result of this electronic structure. This idea is certainly fundamental to the understanding of the way chemicals react.

This chapter describes the last of the four quantum numbers, that concerned with electron spin, and then shows how to use the quantum number rules to predict the electronic configuration of an element. This procedure is readily extended to include ions, although the transition metal ions do represent a special case.

Finally, this chapter examines elemental properties, such as atomic size, ionization energy, electron affinity, and ionic size. The periodic variation in the values of these properties is a strong indication that the idea of atomic electronic configurations is accurate. The chapter closes by discussing how electronic properties are responsible for chemical reactivity.

LEARNING GOALS

1. In addition to reviewing the three quantum numbers discussed in the previous chapter, you should also understand the use of the fourth quantum number, m_s, which is related to the spin magnetic moment of the electron. These quantum numbers and the Pauli exclusion principle are fundamental to the theory of atomic electronic configuration. (Sec. 9.1 & 9.2)

2. In order to understand atomic structure, it is essential to learn the various ways of remembering

the relative energy order of the atomic orbitals. The most obvious result of these relative energies is the shape of the periodic table. (Sec. 9.3)

3. You should be able to depict the electronic configuration of any element or simple monatomic ion by
   (a) writing a set of four quantum number values for each electron in the atom,
   (b) using orbital box diagrams, or
   (c) using spectroscopic notation.
In each of these cases, the notation can be simplified by using the appropriate rare gas configuration. (Sec. 9.4 & 9.5)

4. Be able to use the periodic table to organize the periodic trends in properties such as atomic radius, ionization energy, electron affinity, and ionic size when comparing elements across a period or down a group. (Sec. 9.6)

ARMCHAIR EXERCISE

1. Suppose that a friend of yours is writing a science fiction story involving an alternative universe where the quantum number rules are slightly different from those that are actually known to be true. He proposes to make one small change; the values of m_s can be +1/2, -1/2, and 0. **All other quantum number rules will remain exactly the same.** To help him understand the results of this change and make sure the story is consistent (even though it is only an imaginary universe), answer the following questions.
a. What is the maximum number of electrons allowed in a single orbital?
b. How many elements are possible in the first period, that is, having $n = 1$?
c. How many elements are possible in the second period, that is, having $n = 2$?
d. How many elements are possible in the third period, that is, having $n = 3$?
e. What is the electron configuration of an element having an atomic number of 5? Will this element be a metal or a nonmetal? What is a probable oxidation state for this element?

f. What is the electron configuration of an element having an atomic number of 26? Will this element be a metal or a nonmetal? What is a probable oxidation state for this element?

g. If the element described in part e has the symbol X, and the element described in part f has the symbol Z, give the formula for a compound formed by the combination of these two elements.

CONCEPT TEST

1. Substances weakly repelled by a strong magnetic field are said to be _____ ; metals and other compounds that are weakly attracted by a magnetic field are said to be _____.

2. The Pauli Exclusion Principle states that no two electrons in an atom can have the same set of four

_____.

3. Each atomic orbital can be occupied by no more than _____ (give number) of electrons and these electrons must have the opposite _____.

4. What is the maximum number of electrons possible in an electronic shell having an $n = 4$? _____

5. For each pair, circle the atomic orbital that is lower in energy. (Hint: Compare the $n + \ell$ value)

(a) 4s or 3d   (b) 4p or 5s   (c) 4f or 5d?

6. Substances that maintain their magnetism when withdrawn from a magnetic field are called

_____.

7. Circle any elements in the following list that are p-block elements.

    Cu   Na   Cl   Cr   Al

8. Circle the elements that are d-block elements.

    Cu   Na   Cl   Cr   Al

9. The charge felt by any electron at any distance from the nucleus in an atom is called the _____ _____.

10. According to Hund's rule, when electrons are assigned to different orbitals in the same subshell, the most stable arrangement is that with the maximum number of _____.

11. Which of the following best describes the variation of atomic radii of the main group elements with respect to their position on the periodic table?

a. decreases across a period, increases down a
   group.
b. decreases across a period, decreases down a
   group.
c. increases across a period, increases down a
   group.
d. increases across a period, decreases down a
   group.

12. For each of the following elements, indicate which rare gas symbol would be used if we wished to abbreviate the electronic configuration.

Ag _____   Ti _____

Ba _____   Fe _____

13. Which of the following best describes the variation of first ionization energies of the elements with respect to their position on the periodic table?

a. decreases across a period, increases down a
   group.
b. decreases across a period, decreases down a
   group.
c. increases across a period, increases down a
   group.
d. increases across a period, decreases down a
   group.

14. Electron spin is quantized is such a way that, in an external magnetic field, _____ (give number) orientations of the electron spin are possible.

15. When a cation is formed from an atom, is the size of the cation larger than, smaller than, or the same size as the atom which was ionized?

_____

16. When an anion is formed from an atom, is the size of the anion larger than, smaller than, or the same size as the atom which was ionized?

_____

17. The enthalpy change for the reaction of gaseous ions to form a crystalline lattice is called the

_____.

18. Main group elements form cations with an electron configuration equivalent to that of the nearest _____.

STUDY HINTS

1. The best way to remember the filling order of the atomic orbitals is to understand the relationship between the periodic table and the atomic electronic configurations. As you write electronic configurations, compare them with the periodic table and try to see how the table is simply the result of the filling of the different types of orbitals.

Another way to remember the filling order is to construct the diagram on the right. Simply list the possible orbitals in the order shown, and then draw diagonal lines from upper right to lower left to show the filling order.

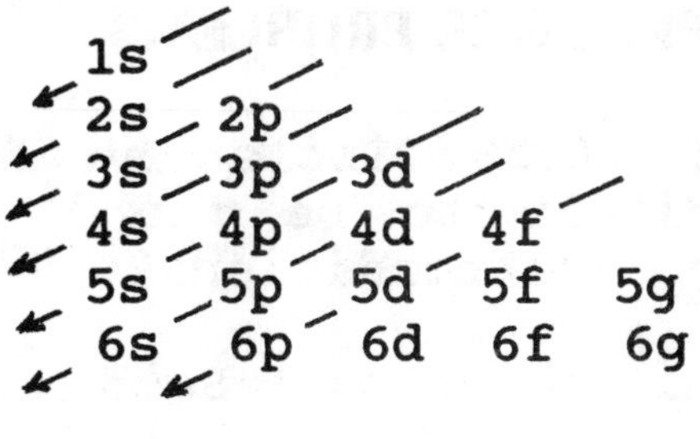

```
1s
2s  2p
3s  3p  3d
4s  4p  4d  4f
5s  5p  5d  5f  5g
6s  6p  6d  6f  6g
```

2. Normally we use a filling order chart, like that in the previous hint, to predict what the periodic table will look like. Actually the best assistance to remembering the electronic configurations is the table itself. With a little practice, you will find that the table itself is the best possible aide to remembering most of the orbital filling orders.

3. Remember that both the periodic chart and the diagram above only give the filling order of the orbitals. For instance, the 4s fills before the 3d, but once both orbitals are filled the 3d is lower in energy than the 4s. That is, the electron configuration of germanium would be written as

$$1s^2 2s^2 2p^6 3s^2 3p^6 3d^{10} 4s^2 4p^2$$

rather than $1s^2 2s^2 2p^6 3s^2 3p^6 4s^2 3d^{10} 4p^2$.

Check to determine which way your instructor wishes you to write the spectroscopic notation.

4. Remember that the law of chemical periodicity is based on atomic numbers, not atomic masses. Throughout most of the periodic chart, the elements seem to be arranged in order of increasing atomic mass, but this is not always true. Can you find three places on the periodic table where the order of the elements is the reverse of that expected from the atomic masses?

5. Be sure that you recognize the difference between electron affinity and ionization energy, especially the sign on the energy change! It always requires energy to ionize an atom, but energy is always released when an electron is added to form a stable anion.

PRACTICE PROBLEMS

1. (L3) Circle the following ions which are <u>not</u> likely to be found in chemical reactions under normal conditions?

Li^+ S^{3-} Sc^{4+} Mg^{2+} Ne^+ Br^-

2. (L3) a. Which element is the first p block element? _____

b. Which element is the
first d block element?   _____

3. (L3) Write electron configurations for each of
these elements belos using both orbital box
diagrams and spectroscopic notation. Also indicate
for each element whether it is paramagnetic or
diamagnetic. You may use rare gas notation in part
c to simplify your answer.

(a) magnesium (atomic number = 12)

(b) phosphorus (atomic number = 15)

(c) cobalt (atomic number = 27).

4. (L4) Arrange the following elements in order of
increasing first ionization energy: Cs, C, K, Li,
and F.

5. (L4) Arrange the following ions in order of
increasing size: O^{2-}, N^{3-}, and F^-.

6. (L3) Use spectroscopic notation to represent the
electron configuration of the following elements.
Indicate for each element whether or not it is
paramagnetic. Use rare gas notation if you wish.

(a) The element zirconium, atomic number = 40, is
found in a number of minerals that have been known
since biblical times, but the pure element wasn't
isolated until 1914.

(b) The radioactive element polonium, atomic number
= 84, was the first element discovered by Mme.
Curie and was named for Poland, her native country.

(c) Iodine, element 53, can cause health problems either by being present or being absent. A small amount of iodine in the diet is necessary to prevent goiter, but iodine vapor is irritating to the eyes and can cause lesions on the skin.

7. (L3) Write down a complete set of four quantum numbers for each of the electrons beyond the nearest rare gas for the elements
(a) praseodymium (atomic number = 59)

(b) scandium (atomic number = 21).

8. (L1) What is the maximum number of electrons in an atom that could have each of the following sets of quantum numbers (assuming all possible values for the other quantum numbers)? If the answer is none, explain why this is so.

(a) $n = 2$, $\ell = 2$, $m_\ell = 0$ _____

(b) $n = 5$, $\ell = 2$ _____

(c) $n = 6$, $\ell = 2$, $m_\ell = 1$, $m_s = +1/2$ _____

9. (L4) The electron configurations of A and B, two unknown elements, are

$$A = \ldots\ 3s^2 3p^5 \qquad\qquad B = \ldots\ 3s^2 3p^2$$

a. Indicate whether each element is a metal, a metalloid, or a nonmetal.

A _____ B _____

b. Which of these is expected to have a smaller atomic radius? _____

144

c. Predict the formula of a likely compound formed only by these two elements.

_____

10. (**L4**) Name each of the following elements based on the information provided.

a. electronic configuration is
$1s^2 2s^2 2p^6 3s^2 3p^6 4s^2 3d^8$ _____

b. the alkali metal with the largest ionization energy _____

c. the element whose +4 ion has the configuration $[Kr]4d^3$ _____

d. the first element (that is the one with the lowest atomic number) that has f electrons in its ground state electronic configuration. _____

11. (**L3**) Use spectroscopic notation (and the appropriate noble gas symbol) to represent the electron configuration for each of the following ions:

a. Ni^{2+}

b. S^{2-}

c. Mo^{3+}

d. Pt^{2+}

PRACTICE PROBLEM SOLUTIONS

1. The following ions do not have a noble gas configuration, and so they are not expected to be found in nature under normal conditions: S^{3-}, Sc^{4+}, and Ne^+.

2. Boron is the first element (i.e. the one with the smallest atomic number) that has p electrons, and scandium is the first d block element.

3. a. magnesium - diamagnetic

(↑↓) (↑↓) (↑↓) (↑↓) (↑↓) (↑↓)
 1s   2s   2p   3s

spectroscopic notation: $1s^2 2s^2 2p^6 3s^2$

b. phosphorus - paramagnetic

(↑↓) (↑↓) (↑↓) (↑↓) (↑↓) (↑↓)   (↑)(↑)(↑)
 1s   2s   2p   3s   3p

spectroscopic notation: $1s^2 2s^2 2p^6 3s^2 3p^3$

c. cobalt - paramagnetic
 [Ar] (↑↓)   (↑↓) (↑↓) (↑)(↑)(↑)
         4s   3d

spectroscopic notation: $1s^2 2s^2 2p^6 3s^2 3p^6 4s^2 3d^7$

4. Cs < K < Li < C < F

5. $F^- < O^{2-} < N^{3-}$

6. a. zirconium [Kr] $5s^2 4d^2$ paramagnetic
or
$1s^2 2s^2 2p^6 3s^2 3p^6 3d^{10} 4s^2 4p^6 5s^2 4d^2$

b. polonium [Xe] $6s^2 4f^{14} 5d^{10} 6p^4$ paramagnetic
or
$1s^2 2s^2 2p^6 3s^2 3p^6 3d^{10} 4s^2 4p^6 4d^{10} 4f^{14} 5s^2 5p^6 5d^{10} 6s^2 6p^4$

c. iodine [Kr] $5s^2 4d^{10} 5p^5$ paramagnetic
or
$1s^2 2s^2 2p^6 3s^2 3p^6 3d^{10} 4s^2 4p^6 4d^{10} 5s^2 5p^5$

7. a.

n	ℓ	m_ℓ	m_s
6	0	0	+1/2
6	0	0	-1/2
4	3	+3	+1/2
4	3	+2	+1/2
4	3	+1	+1/2

Other combinations are possible for the f electrons.

b.

n	ℓ	m_ℓ	m_s
4	0	0	+1/2
4	0	0	−1/2
3	+2	+2	+1/2

Other combinations are possible for the 3d electron.

8. a. none, n and ℓ cannot have the same value.
   b. This corresponds to a set of five 5d orbitals, and so may contain a maximum of 10 electrons.
   c. This corresponds to a single 6d electron.

9. a. A is a nonmetal; B is a metalloid.
   b. A should have the smaller atomic radius.
   c. BA_4

10. a. nickel
    b. lithium
    c. technetium
    d. cerium

11. a. Ni^{2+} $[Ar]3d^8$
    b. S^{2-} $[Ne]3s^2 3p^6$
    c. Mo^{3+} $[Kr]4d^3$
    d. Pt^{2+} $[Xe]4f^{14}5d^8$

PRACTICE TEST (40 Minutes)

1. Give the symbols of all of the elements that in their ground states have (a) three p electrons in their outermost subshell and (b) two d electrons in their outermost subshell.

2. List all of the elements in the first three periods of the periodic table that have two unpaired electrons in their ground state.

3. Write electron configurations for the following elements using both orbital box diagrams and spectroscopic notation: (a) K and (b) As. Indicate for each element whether or not it is paramagnetic.

4. Arrange the following elements in order of decreasing atomic radius: Si, Cl, Mg, Sr, and F.

5. Arrange the following elements and ions in order of decreasing size: Se^{2-}, Kr, Br^-, and As^{3-}.

6. What is the maximum number of electrons that can be identified with each of the following sets of quantum numbers? If the answer is none, explain why this is so.

(a) $n = 3$

(b) $n = 3$, $\ell = 2$, $m_\ell = -2$

(c) $n = 5$, $\ell = 3$, $m_\ell = -3$, $m_s = 0$

7. Name each of the following elements based on the information provided.

a. the electronic configuration is $[Ar]4s^2 3d^8$
b. the element whose +3 ion has the configuration $[Ar]3d^2$
c. the element in Group 5A with the highest first ionization energy.
d. the element in the third period with the greatest number of d electrons.

8. Use spectroscopic notation to represent the electron configuration of the following elements. Indicate for each element whether or not it is paramagnetic. You may use the rare gas notation if you wish.

a. The element palladium, atomic number = 46, is used for dentistry, watch and instrument making, and also as a catalyst.

b. Gallium, atomic number = 31, is one of the few metals that is a liquid at about room temperature. (Can you name another metal that is a liquid?)

c. Rhenium, atomic number = 75, is a very dense element with a high melting point used as a catalyst and also in many different alloys, where it increases corrosion and wear resistance.

148

9. Use Spectroscopic notation to represent the electron configuration for each of the following ions:

a. Zn^{2+} b. P^{3-} c. Nb^{3+}

10. Write down a complete set of four quantum numbers for each of the electrons beyond the nearest rare gas for the elements:
(a) manganese (atomic number = 25) and (b) cadmium (atomic number = 48)

COMMENTS ON ARMCHAIR EXERCISE

a. three   b. three
c. twelve   d. twenty seven
e. $1s^3 2s^2$, metal, +2
f. $1s^3 2s^3 2p^9 3s^3 3p^8$, nonmetal, -1
g. XZ_2

DON'T CONFUSE THIS PROBLEM WITH REALITY!

CONCEPT TEST ANSWERS

1. diamagnetic, paramagnetic
2. quantum numbers
3. two, spin directions
4. 2 x n x n = 32
5. 4s, 4p, & 4f are lowest.
6. ferromagnetic
7. Al and Cl
8. Cu and Cr
9 shielded nuclear charge
10. unpaired electrons
11. a. decreases across a period, increases down a group.
12. Ag _Kr_   Ti _Ar_   Ba _Xe_   Fe _Ar_
13. d. increases across a period, decreases down a group.
14. two
15. smaller
16. larger
17. lattice energy
18. noble (or rare) gas

PRACTICE TEST ANSWERS

1. **(L3)** a. N, P, As, Sb, and Bi
   b. Ti, Zr, and Hf

2. **(L3)** carbon, silicon, and oxygen

3. **(L3)** a. potassium - paramagnetic

   [Ar] (↑)
    4s

spectroscopic notation: $1s^2 2s^2 2p^6 3s^2 3p^6 4s^1$

b. arsenic - paramagnetic

[Ar] (↑↓) (↑↓)(↑↓)(↑↓)(↑↓)(↑↓)   (↑)(↑)(↑)
 4s 3d 4p

spectroscopic notation: $1s^2 2s^2 2p^6 3s^2 3p^6 4s^2 3d^{10} 4p^3$

4. **(L4)** Sr > Mg > Si > Cl > F

5. **(L4)** As^{3-} > Se^{2-} > Br^- > Kr

6. **(L1)** a. The n = 3 set of orbitals may contain 18 electrons.
   b. A single 3d orbital may contain two electrons.
   c. Because m_s = 0, this is not possible.

7. **(L3)** a. nickel b. vanadium
   c. nitrogen d. zinc

8. **(L3)** a. $[Kr]5s^2 4d^8$ - paramagnetic
   b. $[Ar]4s^2 3d^{10} 4p^1$ - paramagnetic
   c. $[Xe]6s^2 4f^{14} 5d^5$ - paramagnetic

9. **(L3)** a. $[Ar]3d^{10}$ b. $[Ne]3s^2 3p^6$ c. $[Kr]4d^2$

10. **(L3)** a. manganese

n	ℓ	m_ℓ	m_s
4	0	0	+1/2

10. a. manganese (cont.)

n	ℓ	m_ℓ	m_s
4	0	0	-1/2
3	+2	+2	+1/2
3	+2	+1	+1/2
3	+2	0	+1/2
3	+2	-1	+1/2
3	+2	-2	+1/2

a. cadmium

n	ℓ	m_ℓ	m_s
5	0	0	+1/2
5	0	0	-1/2
4	+2	+2	+1/2
4	+2	+1	+1/2
4	+2	0	+1/2
4	+2	-1	+1/2
4	+2	-2	+1/2
4	+2	-2	-1/2
4	+2	-1	-1/2
4	+2	0	-1/2
4	+2	+1	-1/2
4	+2	+2	-1/2

CHAPTER 9
BONDING AND MOLECULAR STRUCTURE: <u>FUNDAMENTAL CONCEPTS</u>

CHAPTER OVERVIEW

Chemical bonding is a fundamental chemical process, and there are several theories which attempt to represent this process. Some of these theories are very complex and mathematical; some are quite simple. You must not only understand each of these theories but also recognize when each would be most useful.

In this chapter you encounter two of the simpler theories, the Lewis dot diagrams and Valence Shell Electron Pair Repulsion Theory. The Lewis Theory is one of the oldest theories of chemical bonding which is still widely used; Lewis proposed it before our modern picture of atomic structure was well developed. In the original form, it provided little information about molecular structure, that is, how the atoms are spatially arranged in the molecule. It does, however, show how electrons determine the stability and reactivity of molecules.

As you work with the Lewis theory, remember that the octet rule is just a reflection of the number of electrons that will fit in the s and p orbitals for a single period. As you encounter exceptions, try to understand how they are related to the match between the available electrons and the available spaces in the atomic orbitals. For example, many of the exceptions occur when the presence of d orbitals allows for more than eight electrons in a filled shell.

The second major theory discussed in this chapter is Valence Shell Electron Pair Repulsion Theory. This is based on the idea that electron pairs will distribute themselves on the surface of a spherical central atom in such a way as to be as far apart as possible, that is, to minimize repulsion. This theory does predict the structure of many molecules. Although theoreticians have expressed skepticism about its simplicity, there is no denying that this model summarizes a large amount of information about molecular structure.

This chapter also includes some of the basic terms that will be used to describe chemical bonds, including bond order, bond length, bond polarity, and bond dissociation energy. Pay careful attention to this new vocabulary.

LEARNING GOALS

1. Chemical bonding involves the atomic valence electrons, and so in order to understand bonding it's necessary to know how to identify the valence electrons of each element. The difference between covalent and ionic bonding is also an important idea to master at this stage. (Secs. 9.1 & 9.2)

2. Lewis electron dot structures are one of the oldest methods of describing covalent bonding, but they are still widely used. Learn to draw Lewis dot structures for simple molecules and ions, including being able to identify lone pair and bond-pair electrons. In addition, you should know how to use the Lewis diagrams to represent cases where the octet rule is not followed, including odd-electron compounds, expanded valence compounds, and compounds where the central atom has less than eight valence electrons. (Sec. 9.3)

3. Molecules that have resonance structures are simply a special case of Lewis electron dot representation. Know how to represent these species. (Sec. 9.3)

4. Understand the major terms used to describe a chemical bond, including bond order, bond length, and bond dissociation energy, and understand the relationship between bond order and bond length and energy. When appropriate data is provided, be able to estimate bond energies from enthalpies of formation, or vice versa. (Sec. 9.4)

5. Study carefully the rules in Section 9.4 that show how to determine the polarity of a chemical bond and understand the important role that electronegativity plays in polarity. Understand

153

how to determine the atom formal charge for the atoms in either a molecule or an ion. (Sec. 9.4)

6. The three-dimensional structure of molecules provides important information that may be used for several different purposes, including the determination of chemical reactivity. The Valence Shell Electron Pair Repulsion (VSEPR) theory provides a useful representation of molecular structure. Learn the electron-pair geometry associated with two to six pairs of bonding electrons and use this information to predict the molecular geometry for molecules that have lone pairs as well as those that do not. Based on these predicted structures, you should be able to estimate the approximate bond angles in these molecules. (Sec. 9.5)

7. When electronegativity differences causes a molecule to have a positive and a negative pole, the molecule is said to be a dipole. Understand how to recognize a dipolar molecule by combining information about molecular structure with the electronegativity difference. (Sec. 9.6)

ARMCHAIR EXERCISES

1. Boron trifluoride is an exception to the octet rule because boron doesn't supply enough electrons to complete the octet. As you might expect, the empty place in the octet makes boron trifluoride very reactive, and for this reason it is an important industrial chemical. Suggest some molecules that might react readily with boron trifluoride. HINT: If boron trifluoride lacks a pair of electrons, what molecules might have an extra pair of electrons to donate to the BF_3?

2. The armchair exercise in the previous chapter (pg. 138) suggested that a friend of yours was writing a science fiction story about an imaginary universe, where the spin quantum number could have three different possible values instead of two. What change would this make in the octet rule which is discussed in this chapter. Briefly discuss how bonding would be different in this imaginary universe.

154

CONCEPT TEST

1. The number of valence electrons of each main group element is equal to that element's _____ _____ on the periodic table.

2. When a chemical bond involves the complete transfer of one or more electrons from one atom to another, the resulting bond is said to be _____; when the electrons are shared the bond is said to be _____.

3. Ionic bonds are generally formed when _____ combine with _____. (In each case, give the type of element involved.

4. Covalent bonding generally involves the single type of element known as _____.

5. For a covalently bonded molecule, pairs of valence electrons that are not involved in the bonding are said to be _____.

6. When writing Lewis structures, the central atom is the atom having the lowest _____.

7. In Lewis structures, hydrogen is almost always attached to _____ (how many) other atom(s). Multiple bonds are most often formed by atoms of the following four elements: _____, _____,

_____, and _____.

8. When there is more than one way to correctly draw the Lewis Structure of a molecule, and the actual structure is best represented by a combination of these individuals structure, the situation is called _____.

9. There are several common exceptions to the octet rule. For example, the element _____ is particularly noted for forming compounds in which it shares only three electron pairs and the main group elements beyond _____ may share in more than four pairs.

10. The number of bonding electron pairs shared by two atoms in a molecule is called the _____.

11. _____ is the distance between the nuclei of two bonded atoms.

12. When a series of chemical bonds involving the same pair of atoms is compared, the shortest bond length is expected to be the one that has the highest _____.

13. List these pairs of bonded atoms in order of <u>increasing</u> bond distance: O-F, O-I, and O-Cl.

_____

14. The enthalpy change for breaking a bond in a molecule with the reactants and products in the gas phase under standard conditions is called the

_____.

15. The process of breaking bonds in a molecule is

_____. (Circle the best answer.)

a. always exothermic   b. always endothermic

c. may be exothermic or endothermic depending on

   conditions.

16. A chemical bond that has a positive end and a negative end (that is, electric poles), is called a

_____.

17. _____ is a measure of the

ability of an atom in a molecule to attract

electrons to itself.

18. Arrange the following pairs of bonded atoms in order of increasing bond polarity: C-Cl, Na-Cl, and C-C. For polar bonds, underline the more positive atom.

19. Complete the expression

Atom formal charge = _____

           - the number of lone pair electrons

           - ____ (number of bonding electrons)

20. The oxidation number of an atom is the charge that the atom would have if ALL of its bonds were considered to be completely _____.

21. List the expected electron-pair geometry for each number of electron pairs.

(a) 6 _____

(b) 3 _____

(c) 5 _____

(d) 2 _____

22. The VSEPR model does not apply to those molecules where the central atom is one of the elements known as a(n) _____.

23. The electron pair geometry for a molecule will be the same as the molecular geometry if the central atom has no _____.

24. The _____ is defined as the product of the magnitude of the partial charges in a polar molecule multiplied and the distance of separation for these charges.

STUDY HINTS

1. When no single Lewis structure will adequately represent a molecule, it is sometimes necessary to use a combination of several structures called resonance structures. The actual structure may be thought of as a combination of all of these

resonance forms. Remember that these are alternative pictures of the same molecule, so they only differ in the distribution of electron pairs. All resonance structures of the same molecule must have the same atoms in the same positions, bonded to the same partners.

2. Many students initially have difficulty visualizing three-dimensional structures. If you have this problem, it may be helpful to work with a set of three-dimensional models. Practice by trying to visualize the molecules shown in your textbook. In each case first look at the formula on the page and try to mentally picture how the model will look. Then use the model to see how closely your mental image corresponds to the real thing. Some individuals will find this easier than others, but practice will help anyone to improve this skill.

3. It is important for you to distinguish between the electron-pair geometry and the molecular geometry. If a molecule has no electron lone pairs, the arrangement of the atoms in space must be identical with the arrangement of the electron pairs. For molecules that do have lone pairs, the electron-pair geometry will always be different from the molecular geometry. Even in this case, however, the two are related, since the position of the atoms must be determined by the position of the bonding electron pairs. The best approach is to first determine the electron-pair geometry, then determine how many lone pairs are present, and finally predict the molecular geometry based only on the arrangement of the bonding electron pairs.

4. It is important to remember that the lone pairs in a trigonal bipyramidal electron-pair geometry will be in the equatorial positions, otherwise you will have difficulty predicting the correct molecular geometry for these cases. If you are not sure of the difference between the axial and equatorial positions, use your molecular models to clarify this point.

5. There is one important limitation that you should remember when using the VSEPR Theory. Although it does an excellent job of predicting the

structure of compounds that have a typical metal or a typical nonmetal as the central atom, it is frequently incorrect for compounds that have a transition metal as the central atom.

6. Remember that even though some or all of the chemical bonds in a molecule may be polar, the molecule can still be nonpolar because of the molecular geometry.

PRACTICE PROBLEMS

1. **(L1)** Give the periodic group number and the number of valence electrons for each of the following atoms:

        (a) Si (b) F (c) Al (d) Li (e) Be

group
number   ____ ____ ____ ____ ____

no. of
valence
electrons ____ ____ ____ ____ ____

2. **(L2)** Which of the following atoms could have an expanded valence shell and form compounds with five or more valence electron pairs?

(a) S   (b) Si   (c) N   (d) O   (e) Br

3. **(L2)** Draw Lewis dot structures for the following molecules or ions:
(a) Cl_2 (b) CO_2

(c) NCl_3 (d) GeO

(e) ClO_2^- (f) NO_2^+

4. **(L4)** What is the bond order of each chemical bond in the following molecules.

```
          O                          H   O
          ‖                          |   ‖
(a) H – C – O – H            (b) H – C – C – H
                                     |
                                     H
```

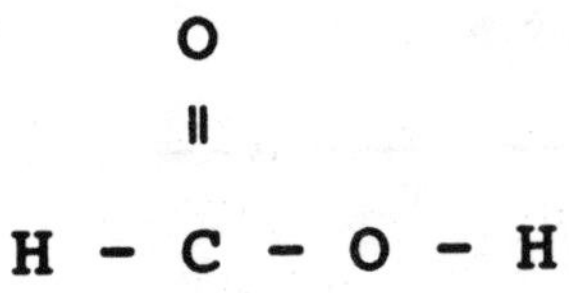

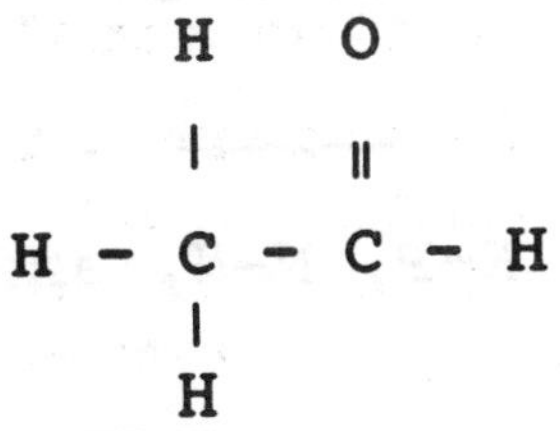

5. **(L3)** The following molecules or ions have two or more resonance structures. Show all of the resonance structures for each species.

(a) SeO_3

(b) NO_3^-

6. **(L5)** The combination of hydrogen and oxygen gases, a very energetic process that can be used to power a rocket, is described by the equation

$$2\ H_2(g)\ +\ O_2(g)\ \rightarrow\ 2\ H_2O(g)$$

Using the bond energies below, estimate the standard heat of combustion per mole of hydrogen gas.

Bond	Bond Energy
H – H	436 kJ/mol
O = O (O_2)	498
H – O	464

7. **(L4)** Arrange the following pairs of bonded atoms in terms of increasing bond length:

(a) N-Cl, N-Br, and N-F

_____ < _____ < _____

(b) N-P, N-N, and N-As

_____ < _____ < _____

8. **(L7)** Arrange the following pairs of bonded elements in order of increasing polarity of the C - X bond:

(a) C - N   (b) C - O   (c) C - I   (d) C - F

_____ < _____ < _____ < _____

9. **(L6)** The column on the left represents general formulas for several different types of simple molecules or ions. The letter M represents the central atom; B represents bonding electron pairs, and E represents lone pairs of electrons. For example, NF_3 would be represented MB_3E, since there are three bonding pairs and one lone pair on the nitrogen. The right hand column lists phrases that describe some of the possible molecular shapes that can exist. Match the general formula on the left with the appropriate molecular shape on the right by placing the letter from the right hand column in the space provided.

_____ MB_2E	a. square planar
_____ MB_5E	b. bent (120°)
_____ MB_3E	c. "T-shaped"
_____ MB_3E_2	d. trigonal pyramidal
_____ MB_2E_3	e. square pyramidal
_____ MB_4E_2	f. linear

162

10. **(L6)** Describe the central atom electron-pair shape and the molecular (or ionic) shape (including the <u>approximate</u> bond angle for bent molecules) for each of the following:

formula	electron-pair shape	molecular or ionic shape

a. BeF_2

b. CCl_4

c. BCl_3

d. OF_2

e. $BrCl_3$

f. BCl_2^-

11. **(L5)** Draw the Lewis structures and determine the atom formal charges for each atom in

(a) hydrogen cyanide, HCN

(b) phosphorus oxychloride, $POCl_3$

PRACTICE PROBLEM SOLUTIONS

1. element	group number	number of valence electrons
Si	4	4
F	7	7
Al	3	3
Li	1	1
Be	2	2

2. Sulfur, silicon, and bromine are each in the third or fourth period and so may have compounds with an expanded octet of electrons. Nitrogen and oxygen are second period elements and so cannot have an expanded octet.

3. a. (1) Since only two atoms are present, there isn't really a central atom.
(2) The total number of valence electrons is the sum of the group numbers, that is, since the group number for chlorine is seven, two times 7 = 14. There are seven valence pairs.
(3) Place a single bond between each pair of atoms.

$$Cl - Cl$$

(4) and (5) The remaining six pairs are distributed around the two chlorines as lone pairs.

$$: \overset{..}{\underset{..}{Cl}} - \overset{..}{\underset{..}{Cl}} :$$

Each chlorine now has an octet, so the structure is complete.

3. b. (1) Carbon has a lower electron affinity than oxygen, so carbon is the central atom.
(2) The total number of valence electrons is four from carbon plus 2 times 6 from oxygen, or 16 electrons.
(3) Place a single bond between each pair of atoms.

$$O - C - O$$

(4) and (5) The remaining 12 electrons are distributed around the terminal oxygens, until each has four electron pairs.

$$: \overset{..}{\underset{..}{O}} - C - \overset{..}{\underset{..}{O}} :$$

(6) The carbon has a deficiency of two electron pairs, and so you must convert one lone pair from each of the oxygens into a double bond between that oxygen and the carbon.

$$: \overset{..}{O} = C = \overset{..}{O} :$$

164

Each atom now has a share in four electron pairs.

3. c. (1) Nitrogen is the central atom, since it has the lower electron affinity.
(2) The total number of valence electrons is 26. There are 13 valence pairs.
(3) Place a single bond between each pair of atoms.

$$Cl - N - Cl$$
$$|$$
$$Cl$$

(4) and (5) Ten pairs of electrons remain, but only nine are required to provide an octet around each chlorine. The remaining pair is used to serve as a lone pair that will complete the octet for the nitrogen atom.

```
   ..      ..     ..
  :Cl  -   N   -  Cl:
   ..      |      ..
         : C :
           ..
```

3. d. (1) Since only two atoms are present, there isn't really a central atom.
(2) The total number of valence electrons is the sum of the group numbers, that is, 6 + 4 = 10, and there are five valence pairs.
(3) Place a single bond between each pair of atoms.

$$Ge - O$$

(4) and (5) The remaining four pairs are distributed around the two atoms as lone pairs.

```
      ..      ..
  : Ge : O :
```

(6) Both the germanium and the oxygen are deficient by one pair of electrons. To correct this, move one lone pair from each element into a bonding pair, making a triple bond between the two atoms.

$$: Ge ::: O :$$

3. e. The chlorine is the central atom, and there are twenty valence electrons. First, insert single

bonds between each pair of atoms.

$$O - Cl - O$$

Of the remaining 16 electrons, only 12 are required around the "terminal" oxygens to complete their octets. The last two pairs of electrons are placed on the central chlorine atom, to complete its octet.

Thus, the structure is

$$[: \overset{..}{\underset{..}{O}} - \overset{..}{\underset{..}{Cl}} - \overset{..}{\underset{..}{O}} :]^-$$

3. f. The nitrogen is the central atom, and there are sixteen valence electrons. First, insert single bonds between each pair of atoms.

$$O - N - O$$

Placing the remaining six electron pairs around the terminal oxygen atoms completes their octets, but the nitrogen needs two pairs of electrons.

$$[: \overset{..}{\underset{..}{O}} - N - \overset{..}{\underset{..}{O}} :]^+$$

Transfer one of the lone pairs from each oxygen to the bond between the nitrogen and the oxygen. This provides the electrons needed by the nitrogen.

$$[: \overset{..}{O} = N = \overset{..}{O} :]^+$$

This ion has two double bonds.

4. Except where indicated, the bond order of each bond is one. [2]   [2]

(a)
$$H - O - \overset{\overset{\displaystyle O}{\|}}{C} - H$$

(b)
$$H - \overset{\overset{\displaystyle H}{|}}{\underset{\underset{\displaystyle H}{|}}{C}} - \overset{\overset{\displaystyle O}{\|}}{C} - H$$

5. a. The selenium is the central atom, and the number of valence electron is 24. Begin by placing one bond between each of the atoms.

$$O - \overset{}{\underset{\overset{|}{O}}{Se}} - O$$

166

If the remaining 18 electrons (9 pairs) are distributed along the "terminal" oxygens, there are just enough electrons to provide an octet for each oxygen. The selenium lacks one pair of electrons, and normally we would transfer a pair from one of the atoms bonded to the selenium. Since all three oxygens are identical, why does only <u>one</u> form the double bond with selenium? This is the reason why resonance forms are required, to make a double bond equally possible along each of the selenium-oxygen bonds.

$$:\overset{..}{O} - Se = \overset{..}{O}: \quad \leftrightarrow \quad :\overset{..}{O} - Se - \overset{..}{O}: \quad \leftrightarrow \quad O = Se - \overset{..}{O}:$$
$$\quad\quad | \quad\quad\quad\quad\quad\quad\quad\quad || \quad\quad\quad\quad\quad\quad | $$
$$\quad\quad :\overset{..}{O}: \quad\quad\quad\quad\quad\quad :O: \quad\quad\quad\quad\quad\quad :\overset{..}{O}:$$

b. Here, nitrogen is the central atom, and the number of valence electron is also 24. Begin, as usual, by placing one bond between each of the atoms.

$$O - N - O$$
$$\quad\quad |$$
$$\quad\quad O$$

As in the previous case, the remaining 18 electrons (9 pairs) are just enough electrons to provide an octet for each "terminal" oxygen. To provide the required pair for the nitrogen, it's necessary to transfer one pair from one of the atoms bonded to the nitrogen. Since all three oxygens are identical, this is another case where resonance forms are required.

$$[:\overset{..}{O} - N = \overset{..}{O}:]^- \leftrightarrow [:\overset{..}{O} - N - \overset{..}{O}]^- \leftrightarrow [O = N - \overset{..}{O}:]^-$$
$$\quad\quad | \quad\quad\quad\quad\quad\quad\quad || \quad\quad\quad\quad\quad\quad | $$
$$\quad\quad :\overset{..}{O}: \quad\quad\quad\quad\quad\quad :O: \quad\quad\quad\quad\quad\quad :O:$$

6. The balanced equation for the oxidation of one mole of $H_2(g)$ is

$$2\ H_2(g) \ + \ O_2(g) \ \rightarrow \ 2\ H_2O(g)$$

Set up a data table of bonds broken and bonds formed.

Bonds Broken	Bond Energy	Number Broken	Total Energy
H – H	436 kJ/mol	2 moles	872 kJ
O = O	498	1	498

Total Energy used to break these bonds = 1370 kJ

Bonds Formed	Bond Energy	Number formed	Total Energy
H – O	464 kJ/mol	4 moles	1856 kJ

Total Energy released forming these bonds =1856 kJ

Net Energy Change = Energy Used - Energy Released

$$= 1370 \text{ kJ} - 1856 \text{ kJ}$$

Net Energy Change = -486 kJ

(Notice the change is negative since there is a net release of energy.)

Since this is the energy change for two moles of hydrogen gas, divide by two to obtain the enthalpy of combustion per mole of $H_2(g)$.

Standard Enthalpy of Combustion = -243 kJ/mol

7. a. Since the bond order is the same in all three cases, and all three bonds involve nitrogen, the only variable quantity is the halogen. The halogen atoms increase in size in the order F<Cl<Br, and so the order of the bond lengths is

N-F < N-Cl < N-Br.

b. In this case the nitrogen atom and the bond order are constant, so the bond length is determined by the other bonded atoms, which (from the position on the periodic table) increase in size in the order N < P < As. Thus the order of bond length is

N-N < N-P < N-As

8. Based on the electronegativity difference, the order is

(least polar) C - I < C - N < C - O < C - F (most)

Of course, depending on the molecular structure, any of these bonds may be in a molecule which in nonpolar overall.

9. MB_2E - b. bent (120°)
      MB_5E - e. square pyramidal
      MB_3E - d. trigonal pyramidal
      MB_3E_2 - c. "T-shaped"
      MB_2E_3 - f. linear
      MB_4E_2 - a. square planar

10.

formula	electron-pair shape	molecular or ionic shape
a. BeF_2	linear	linear
b. CCl_4	tetrahedron	tetrahedral
c. BCl_3	trigonal planar	trigonal planar
d. OF_2	tetrahedron	bent (~109.5°)
e. $BrCl_3$	trigonal bipyramid	"T-shaped"
f. BCl_2^-	trigonal planar	bent (~120°)

11. a. The Lewis structure for HCN is

$$H - C \equiv N :$$

Using the equation
    formal charge = group number
                     - no. of unshared electrons
                        -1/2(no. of shared electrons)
For hydrogen
    formal charge = $1 - 0 - 1/2(2) = 0$
For carbon
    formal charge = $4 - 0 - 1/2(8) = 0$
For nitrogen
    formal charge = $5 - 2 - 1/2(6) = 0$

b. The Lewis structure for $POCl_3$ is

$$: \overset{\textstyle ..}{\underset{\textstyle ..}{O}} :$$

$$: \overset{..}{\underset{..}{Cl}} - \overset{\displaystyle \|}{\underset{\displaystyle |}{P}} - \overset{..}{\underset{..}{Cl}} :$$

$$: \overset{..}{\underset{..}{Cl}} :$$

Using the same equation

    formal charge = group number
                 - no. of unshared electrons
                  -1/2 (no. of shared electrons)

For oxygen
    formal charge = 6 - 4 - 1/2(4) = 0
For phosphorus
    formal charge = 5 - 0 - 1/2(10) = 0
For chlorine (notice that all three are identical)
    formal charge = 7 - 6 - 1/2(2) = 0

PRACTICE TEST (50 min.)

1. Draw Lewis structures for the following molecules or ions:

(a) O_2 (b) PH_3 (c) $AsCl_3$

(d) CCl_4 (e) SCl_2 (f) ClO_3^-

(g) CN^- (h) H_2CS

2. The following ions have two or more resonance structures. Show all of the resonance structures for each species.

(a) PO_3^- (b) NO_2^-

3. Determine the atom formal charges for each atom in the following species.

(a) OCS   (b) NF_3 (c) SCN^-

4. Arrange the following pairs of bonded atoms in terms of increasing bond length:

(a) C=C, C=O, and C=N   (b) C=S, C=Se, and C=O

170

5. What is the bond order of each chemical bond in the following molecules.

(a) $S = C = O$

(b)
```
       H        O    H
       |        ||   |
  H -  C  - O - C  - C  - H
       |             |
       H             H
```

(c) $H - C \equiv N$

6. Ethylene, C_2H_4, is one of the major starting materials used by the petrochemical industry. One way to produce it might be from ethane, C_2H_6, using the reaction

```
     H   H                          H        H
     |   |                           \      /
H -  C - C - H    →    H_2(g)  +      C  =  C    (g)
     |   |                           /      \
     H   H                          H        H
```

Using the table of bond energies provided, estimate the enthalpy change for this reaction. Does it seem likely that this will be a commercially practical way to produce ethylene? Why?

Bond	Bond Energy
C = C	611 kJ/mol
C - C	347
C - H	414
H - H	436

7. Identify each species that has a dipole moment and, where possible, indicate the direction of the net dipole by drawing an arrow with the arrowhead pointing to the negative end of the bond dipole. (Hint: Determine the structure with VSEPR Theory.)

a. O_2

b. OF_2

c. CCl_4

d. I_3^-

e. PF_3

f. CS_2

8. Describe the central atom electron-pair shape and the molecular (or ionic) shape (including the approximate bond angle for bent molecules) for each

of the following:

a. $SeCl_2$

b. $POCl_3$

c. H_2CS

d. SF_6

e. PF_5

f. NF_3

g. XeF_2

h. IF_4^+

i. $XeOF_4$

j. PI_3

k. TeF_5^-

l. SF_4

COMMENTS ON ARMCHAIR EXERCISES

1. Molecules with a lone pair of electrons would be obvious candidates to react with boron trifluoride. Some examples that you are already familiar with are water and ammonia. In fact, both of these compounds will react rather rapidly with boron trifluoride to form solid products.

2. In the exercise in the previous chapter, you found that the hypothetical change in the quantum number rules would allow for a maximum of three electrons to be placed in a single orbital of any type. This means that, when filled, the combination of s and p orbitals would accommodate twelve electrons. What effect might this have on the structure of the compounds in this imaginary universe?

CONCEPT TEST ANSWERS

1. group number
2. ionic, covalent
3. metals, nonmetals
4. nonmetals
5. nonbonding or lone pairs
6. electron affinity or electronegativity
7. one, carbon, nitrogen, oxygen, and sulfur
8. resonance
9. boron, neon
10. bond order
11. bond length
12. bond order
13. O-F < O-Cl < O-I

14. bond dissociation energy
15. a. always endothermic
16. polar bond
17. electronegativity
18. C-C < $\underline{C}$-Cl < $\underline{Na}$-Cl
19. group number, 1/2
20. ionic

21. (a) octahedral   (b) trigonal planar
    (c) trigonal bipyramidal   (d) linear
22. transition metal atom or ion
23. lone pairs   24. dipole moment

PRACTICE TEST ANSWERS

1. (L2) a. 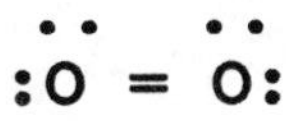 :Ö = Ö:

         ..
b. 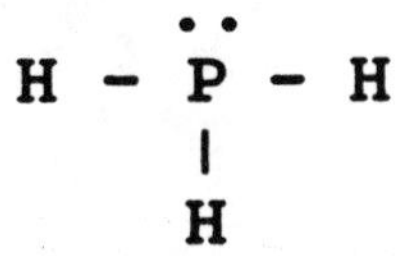 H – P̈ – H
           |
           H

c.
     ..   ..
: Cl̈ – Äs – Cl̈ :
          |
       :Cl̈:
       ..

d.
        ..
      :Cl̈:
     .. | ..
   :Cl̈ – C – Cl̈:
     .. | ..
       :Cl̈:
       ..

e.
   .. .. ..
: Cl̈ – S̈ – Cl̈ :
   .. .. ..

f. [:Ö – Cl̈ – Ö:]⁻
     .. | ..
       :Ö:
       ..

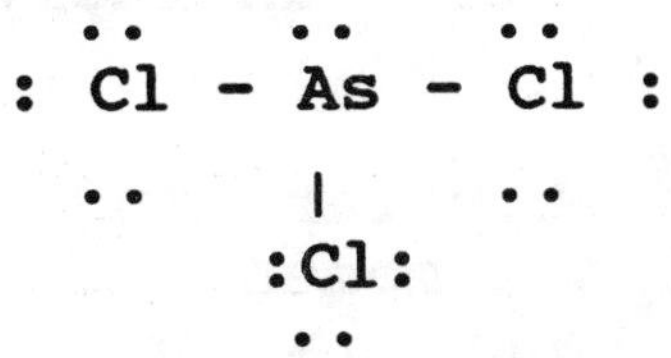

g. [:C ≡ N:]⁻

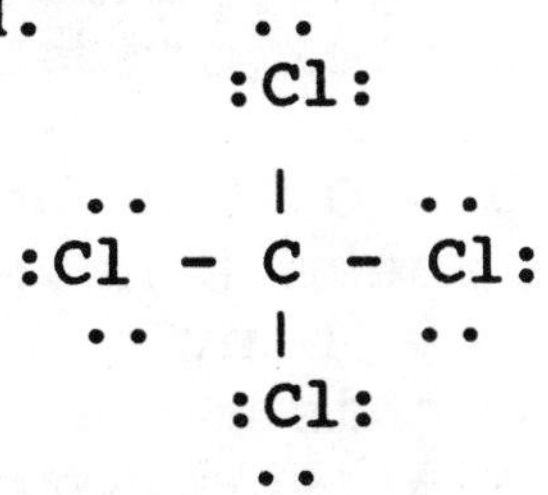

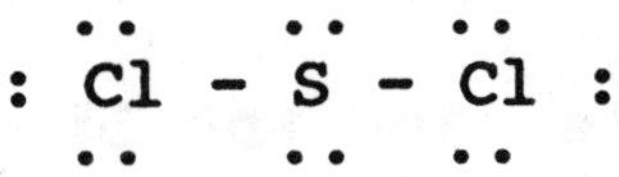

h. H – C – H
          ‖
       : S̈ :

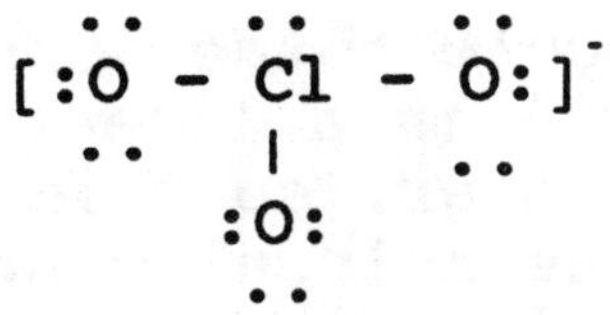

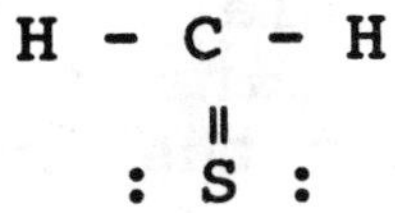

2. (L3)

a. 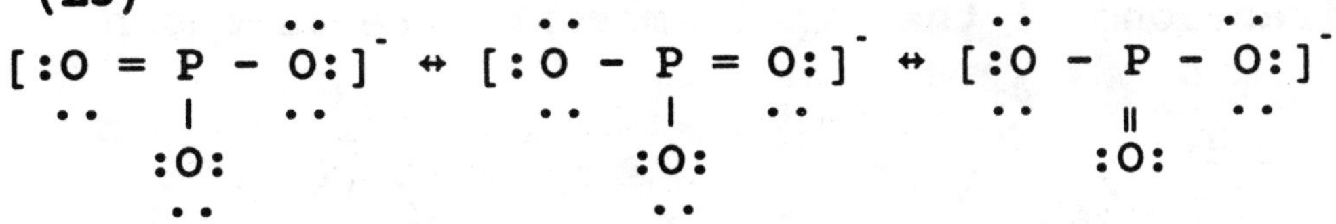 [:Ö = P – Ö:]⁻ ↔ [:Ö – P = Ö:]⁻ ↔ [:Ö – P – Ö:]⁻
   .. | ..   .. | ..   .. ‖ ..
     :Ö: :Ö: :Ö:
     ..   ..

b.  [:Ö = N̈ – Ö:]⁻   ↔   [:Ö – N̈ = Ö:]⁻
         ..   ..

3. (L5) a. sulfur 0, oxygen 0, carbon 0
     b. nitrogen 0, fluorine 0
     c. sulfur –1, carbon 0, nitrogen 0

4. (**L4**) a. C=O < C=N < C=C   b. C=O < C=S < C=Se

5. (**L4**)   All bond orders are one, except those indicated, designated as two or three with bold type and arrows.   **[2]**

$$
\text{a. } S = C = O \quad \text{b. } H - \overset{\displaystyle H}{\underset{\displaystyle H}{\overset{|}{\underset{|}{C}}}} - O - \overset{\displaystyle O}{\overset{\parallel}{C}} - \overset{\displaystyle H}{\underset{\displaystyle H}{\overset{|}{\underset{|}{C}}}} - H \quad \text{c. } H - C \equiv N
$$

6. (**L4**) The calculated $\Delta H^{\circ} = +128$ kJ. Since this reaction is highly endothermic, it's unlikely that it will be useful commercially.

7. (**L7**) a. O = O The bond isn't polar, and the molecule doesn't have a dipole moment.
    b. OF_2 is a bent molecule, and the bonds are polar, so this molecule has a dipole moment.
    c. CCl_4 is a tetrahedral molecule. The bonds are polar, but the symmetrical structure causes the molecule to be nonpolar.
    d. I_3^- is a linear molecule, and the bonds are not polar, so this is also a nonpolar molecule.
    e. The molecular structure of PF_3 is a trigonal pyramid, with a lone pair at one position in the tetrahedron of electron pairs. This will cause the molecule to be polar.
    f. CS_2 is linear (like CO_2). The bonds are polar, but the molecule is not.

For the two polar molecules, OF_2 and PF_3, the directions of the dipole moments are as shown.

8. (**L6**)

formula	electron-pair shape	molecular or ionic shape
a. $SeCl_2$	tetrahedron	bent (109.5°)
b. $POCl_3$	tetrahedron	tetrahedron

174

c. H_2CS	trigonal planar	trigonal planar

8. (cont.)

formula	electron-pair shape	molecular or ionic shape
d. SF_6	octahedron	octahedron
e. PF_5	trigonal bipyramid	trigonal bipyramid
f. NF_3	tetrahedron	trigonal pyramid
g. XeF_2	trigonal bipyramid	linear
h. IF_4^+	trigonal bipyramid	"see-saw" or distorted tetrahedron
i. $XeOF_4$	octahedron	square pyramid
j. PI_3	tetrahedron	trigonal pyramid
k. TeF_5^-	octahedron	square pyramid
l. SF_4	trigonal bipyramid	distorted tetrahedron

CHAPTER 10
BONDING AND MOLECULAR STRUCTURE: ORBITAL HYBRIDIZATION, MOLECULAR ORBITALS AND METALLIC BONDING

CHAPTER OVERVIEW

This chapter introduces two of the more powerful theories of chemical bonding. Unlike the theories in the previous chapter, a rigorous treatment of the theories would be very mathematical, and so this presentation is only qualitative. The bonding theories in this chapter are, however, quite different from those discussed previously.

Historically, the valence bond theory is one of the most important theories of chemical bonding. For almost four decades, starting in the 1940s, it provided the most widely used description of chemical bonding, and the terminology from the theory is still widely used. Particularly in the study of organic chemistry, bonding is still often described by the type of hybridization. Thus, even though the current research directions in chemistry may be somewhat better understood in terms of molecular orbital theory, a knowledge of the valence bond theory is still important.

The molecular orbital theory is currently the most important of the general purpose theories of chemical bonding. It is widely used in many different fields of chemistry and is especially useful for understanding the behavior of metals, insulators and semiconductors.

LEARNING GOALS

1. Understand how to apply the valence bond theory to a variety of simple molecules and identify the type of hybridization and the molecular structure in each case. You should also understand the difference between pi and sigma bonds, as well as recognizing the role that pi bonds play in the formation of cis and trans isomers. (Sec. 10.1)

176

2. Know how to draw molecular orbital electron configurations for simple homonuclear and heteronuclear molecules or ions, as well as understand the special perspective that MO Theory offers for understanding resonance. Based on the molecular orbital diagram, you should be able to determine the bond order, which offers a good measure of stability. (Sec. 10.2)

3. The molecular orbital model is especially useful for explaining the bonding differences among insulators, semiconductors, and conductors. It also serves as the basis for the band theory of metallic bonding. Be able to explain each of these types of behavior in general terms using the molecular orbital theory. (Sec. 10.3)

ARMCHAIR EXERCISES

1. Compare the complexity of the molecules discussed in the valence bond section with those found in the molecular orbital section. Does this reflect what you know about these two theories?

2. One of the interesting tests of chemical bonding theories is to attempt to understand why one element in a period forms a certain type of compound but another element in that same period does not. For example, PCl_5 is known but NCl_5 is not. Why might this be?

3. The $(Cl_5Ru)_2O$ molecule consists of two octahedral ruthenium groups connected through a shared oxygen. The connection from ruthenium to oxygen to the other ruthenium is linear. What does this tell you about the hybridization on the oxygen atom? Can you suggest any reason why this might be true?

4. The fuel for the burners that you use in the laboratory is probably a simple hydrocarbon, like methane or propane. When the burner is well adjusted, the flame has a blue color due to emission of light from high energy C_2 molecules. Can you use molecular orbital theory to describe this molecule?

5. As noted in the textbook, when small amounts of

aluminum are added to silicon (a process called doping), the resulting electron holes can act as acceptors, producing a p-type semiconductor. Can you suggest any other elements that might act like aluminum if they were used to dope silicon?

CONCEPT TEST

1. A chemical bond that has the greatest electron density <u>along the axis</u> of the bond is called a(n) _____ bond; a bond formed by the sideways overlap of p orbitals is a(n)_____ bond.

2. When forming hybrid orbitals, the total number of hybrids formed must always equal _____ _____.

3. Complete the following table.

Hybrid Orbital Set	Geometry of Hybrid Orbital Set
sp	_____
sp^3	_____
sp^3d^2	_____

4. In order for a pi bond to form, there must be _____ on the atom where the hybridization has occurred.

5. Because the pi bonding obstructs free rotation, some molecules with a multiple bond may exist in cis and trans forms called _____.

6. The molecular orbital theory is based on combining pure atomic orbitals from each atom in a molecule to form molecular orbitals that are

_____ over several atoms or even the entire molecule.

7. For each type of hybrid orbital listed on the left below, indicate how many of each type of pure atomic orbital is combined to form the hybrids and the number of orbitals in a set of hybrids.

hybrid type	no. of s orbitals	no. of p orbitals	no. of d orbitals	total hybrids
sp^2	_____	_____	_____	_____
sp^3d	_____	_____	_____	_____

8. The first principle of molecular orbital theory is that the total number of molecular orbitals formed must always equal _____ _____.

9. _____ molecular orbitals are lower in energy than the parent orbitals; _____ orbitals are higher in energy than the parent orbitals.

10. Atomic orbitals combine to form molecular orbitals most effectively when the atomic orbitals are of similar _____.

11. Molecules formed from two identical atoms are said to be _____.

12. For molecular orbital theory, it is convenient to calculate bond order as one-half times the result when the number of electrons in _____ molecular orbitals is subtracted from the number of electrons in _____ molecular orbitals.

13. If the molecular orbital theory is used to

describe a metal, the resulting picture consists of a very large number of molecular orbitals, which are closely spaced within a range of energy values and are delocalized over the atoms of the metal. This is called the _____ theory of metals.

14. The highest occupied band of molecular orbitals in a metal, called the _____ band, is only partially filled.

15. Substances which have a narrow energy gap between the valence band and the conductance band are called _____.

16. In a metal, the highest filled level at absolute zero is called the _____.

STUDY HINTS

1. If you have difficulty determining how many atomic orbitals are actually hybridized, you may wish to use the VSEPR Theory to determine the structure. Then the corresponding type of hybridization becomes obvious if you have memorized the relationships between structure and type of hybridization.

2. Students sometimes make the job of determining the number of sigma and pi bonds much too complicated. You can count on the fact that the first pair of electrons bonding two atoms together will always be sigma. Each additional pair of bonding electrons will be a pi bond. For example, consider the formaldehyde molecule

$$\overset{\displaystyle O}{\underset{\displaystyle H - C - H}{\|}}$$

It has one sigma bond between each carbon and hydrogen. The carbon-oxygen double bond consists of two electron pairs, one of which must be sigma and the other pi.

180

3. Even though the molecular orbital theory may be new to you, don't forget that many of the rules you learned previously, such as Hund's rule and the Pauli principle, still hold true. These rules work here in much the same way that you learned when you did electronic configurations of atoms. For example, you can still depend upon the fact that each orbital, regardless of whether it is a pure atomic orbital, a hybrid, or a molecular orbital, may contain a maximum of two electrons.

4. The answers in this chapter list the orbitals in the order that you would predict based on Figure 10-22. Remember that this energy order for the molecular orbitals is not strictly true for all molecules. It does, however, lead to the correct predictions of magnetism and bond order in the cases that you will encounter.

PRACTICE PROBLEMS

1. (L1) Tell what hybrid orbital set is used by the underlined atom in each of the following molecules or ions:

a. $\underline{N}F_3$ _____   b. $\underline{Si}F_4$ _____

c. $Cl_2\underline{C}=O$ _____   d. $\underline{Al}Cl_4^-$ _____

e. $F\underline{C}N$ _____   f. $\underline{P}Cl_3$ _____

g. $F\underline{N}NF$ _____   h. $O\underline{C}Se$ _____

2. (L1) Tell what hybrid orbital set is used by the oxygen atom in each molecule.

a. O_3 _____   b. CH_3-OH _____   c. FOOF _____

3. (L1) Tell what hybrid orbital set is used by the underlined atom in each molecule or ion:

a. $\underline{S}F_4$ _____   b. $\underline{As}Cl_5$ _____

c. $\underline{S}F_6$ _____   d. $\underline{Cl}F_4^+$ _____

e. $\underline{Al}Cl_6^{3-}$ _____   f. $\underline{Br}Cl_3$ _____

4. (L2) Determine the bond order and molecular orbital electron configuration for each of the following diatomic molecules or ions:

Bond Order	Molecular Orbital Electron Configuration

a. Ne_2 _____ _____

b. Na_2 _____ _____

c. F_2^+ _____ _____

d. Al_2 _____ _____

e. Be_2^+ _____ _____

f. OF _____ _____

g. O_2^{2-} _____ _____

5. (L2) Which of the species in the previous question are diamagnetic?

6. (L2) Based on the bond order predicted by Molecular Orbital Theory, arrange the following species in order of increasing bond length: B_2^+, B_2, and B_2^-. What if you were asked to arrange these species in order of increasing dissociation energy?

7. (L2) 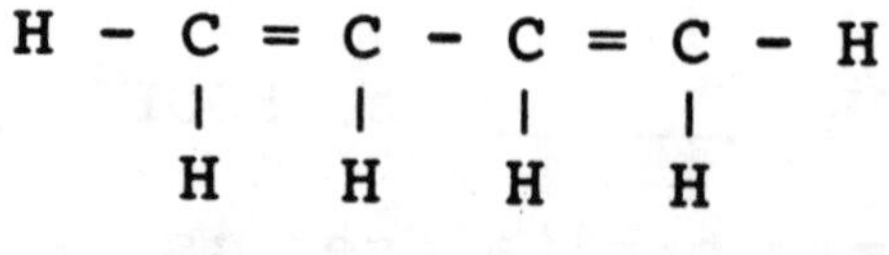 H - C = C - C = C - H
```
           |   |   |   |
           H   H   H   H
```

Above is a compound called butadiene. What is the hybridization on each carbon atom in this molecule? Compare the length of the various carbon-carbon bonds in this molecule.

8. (**L3**) Consider three substances that will be designated A, B, and C. One of these substances is an insulator, one is a conductor and one is a semiconductor. If the band gap in A is 520 kJ/mole, the band gap in B is 105 kJ/mole, and the band gap in C is 7 kJ/mole, assign each of these substances to the most likely category.

PRACTICE PROBLEM SOLUTIONS

1a.

```
      ..     ..     ..
    : F  -   N  -   F :
      ..     |      ..
            : F :
             ..
```

The electron dot structure of NF_3 shows that there are four electron pairs (one lone pair and three bond pairs) that will be at the corners of a tetrahedron. The three bond pairs give the molecule a trigonal pyramidal structure.

As shown in the diagram below, the s orbital and the three p orbitals on the nitrogen atom are hybridized to produce four orbitals, one of which contains a lone pair and the other three each contain one electron that is available for bonding.

```
E     2p   ↑   ↑   ↑              four sp³ hybrid orbitals
N
E  ↑
R                      →      ↑↓      ↑    ↑    ↑
G                            lone    sigma bond
Y     2s      ↑↓             pair    electrons
         isolated N atom
```

The hybridization on the nitrogen is sp^3.

2b.

```
                  ..
                 : F :
            ..    |     ..
          : F  -  Si  -  F :
            ..    |     ..
                 : F :
                  ..
```

The electron dot structure shows that there

183

are four bond pairs at the corners of a tetrahedron, giving a tetrahedral structure.

As shown in the diagram below, the s orbital and the three p orbitals on the silicon atom combine to produce four hybrids, each of which contains one electron that's available for bonding.

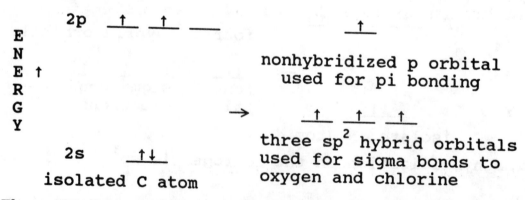

E
N
E ↑
R
G
Y

3p ↑ ↑ __

3s ↑↓

isolated Si atom

→

four sp³ hybrid orbitals

↑ ↑ ↑ ↑

sigma bond electrons

The hybridization on the silicon is sp³.

2c.

: O :
      ‖
: Cl - C - Cl :

The Lewis dot structure for this molecule indicates that there is a double bond from the carbon atom to the oxygen. Since this requires an nonhybridized p orbital on the carbon atom, the hybridization must be as shown below.

E
N
E ↑
R
G
Y

2p ↑ ↑ __

2s ↑↓

isolated C atom

↑

nonhybridized p orbital
used for pi bonding

→

↑ ↑ ↑

three sp² hybrid orbitals
used for sigma bonds to
oxygen and chlorine

Thus the hybridization on the carbon atom is sp².

2d. The electron dot structure shows four sigma bonds connecting the aluminum to the four chlorines. This indicates that the required hybridization would be sp³.

184

2e. The Lewis dot structure is : $\overset{\displaystyle ..}{F} - C \equiv N:$
$\underset{\displaystyle ..}{}$

The hybridization on the carbon must allow two pi bonds, so two p orbitals are not hybridized. The remaining p and s combine to form two sp hybrids.

2f. This molecule resembles NCl_3, with three sigma bonds and a lone pair. All three p orbitals and the one s combine to form a set of four hybrid sp^3 orbitals, one of which is a lone pair.

2g. The Lewis dot structure is : $\overset{\displaystyle ..}{F} - \overset{\displaystyle ..}{N} = \overset{\displaystyle ..}{N} - \overset{\displaystyle ..}{F}:$

There are two sigma bonds, one pi bond, and a lone pair on the nitrogen atom. This means that the hybridization on N must be sp^2.

2h. The Lewis dot structure is : $O = C = \overset{\displaystyle ..}{Se}:$

There are two sigma and two pi bonds on the carbon, so only one p orbital participates in the hybridization. Thus the hybridization must be sp.

2. a. sp^2 b. sp^3 c. sp^3

3a. The Lewis electron dot structure indicates that sulfur has an expanded set of valence electrons

$$
\begin{array}{ccc}
 & :\overset{\displaystyle ..}{F}: & \\
 & | & \\
:\overset{\displaystyle ..}{\underset{\displaystyle ..}{F}} - & \overset{\displaystyle .}{S} - & \overset{\displaystyle ..}{\underset{\displaystyle ..}{F}}: \\
 & | & \\
 & :\underset{\displaystyle ..}{F}: &
\end{array}
$$

This requires a set of sp^3d hybrids.

The diagram on the next page shows the hybridization on the sulfur atom in SF_4 to form a set of five sp^3d hybrid orbitals.

```
3d  __  __  __  __  __          __  __  __  __
```

E
N
E ↑ 3p ↑↓ ↑ ↑   → ↑↓ ↑ ↑ ↑ ↑
R
G
Y   3s ↑↓

4 nonhybridized d
orbitals

five sp^3d hybridized
orbitals for four sigma
bonds and one lone pair

isolated S atom

3b. The Lewis dot structure for $AsCl_5$ shows an expanded valence shell, with five sigma bonds to the five chlorines.
   This means that the hybridization must be sp^3d.

3c. SF_6 has 24 electron pairs, which produces a Lewis dot diagram having six sigma bonds from the sulfur to the six fluorine atoms. To form these six bonds, the sulfur must have an expanded valence shell, using two of the d orbitals for hybridization.
   The hybridization must be sp^3d^2.

3d. The ClF_4^+ ion has 17 electron pairs. When they are distributed in a Lewis dot structure, the result is four sigma bonds from chlorine to fluorine and one lone pair.
   The hybridization is sp^3d.

3e. This molecule has 24 valence electron pairs, and there are six sigma bonds in the Lewis structure. To produce six sigma bonds on the aluminum, the valence shell must be expanded by the addition of two d orbitals.
   The hybridization is sp^3d^2.

3f. The $BrCl_3$ molecule has 14 valence electron pairs. In the Lewis structure, there are two lone pairs and three sigma bonding pairs on the bromine. This requires five hybrid orbitals.
   The hybridization on the bromine is sp^3d.

186

4. | | Bond Order | Molecular Orbital Electron Configuration |

a. Ne_2 0 $(\sigma_{2s})^2(\sigma^*_{2s})^2(\pi_{2p})^4(\sigma_{2p})^2(\pi^*_{2p})^4(\sigma^*_{2p})^2$

b. Na_2 1 $(\sigma_{3s})^2$

c. F_2^+ 1 1/2 $(\sigma_{2s})^2(\sigma^*_{2s})^2(\pi_{2p})^4(\sigma_{2p})^2(\pi^*_{2p})^3$

d. Al_2 1 $(\sigma_{3s})^2(\sigma^*_{3s})^2(\pi_{3p})^2$

e. Be_2^+ 1/2 $(\sigma_{2s})^2(\sigma^*_{2s})^1$

f. OF 1 1/2 $(\sigma_{2s})^2(\sigma^*_{2s})^2(\pi_{2p})^4(\sigma_{2p})^2(\pi^*_{2p})^3$

g. O_2^{2-} 1 $(\sigma_{2s})^2(\sigma^*_{2s})^2(\pi_{2p})^4(\sigma_{2p})^2(\pi^*_{2p})^4$

5. The diamagnetic molecules are Na_2 and O_2^{2-}. Ne_2 would be diamagnetic, but it doesn't exist.

6. In a list of similar molecules, the bond length will become larger as the bond order decreases. The expected order is B_2^- (1 1/2) < B_2 (1) < B_2^+ (1/2), with the bond order in parentheses after each species. The dissociation energy will become smaller as the bond order decreases, so the order bond energy is the reverse of that for bond length.

7. Each carbon atom has the same hybridization, sp^2, and each of the carbon-carbon bonds must be the same length because what appears to be alternating double and single bonds is actually the same bond order in each case because of resonance.

8. The largest band gap is usually associated with an insulator, and so A is probably an insulator. The smallest band gap, for C, indicates a conductor, and the intermediate value for B, suggests that this material is a semiconductor.

PRACTICE TEST (45 min.)

1. What hybrid orbitals are used by each nitrogen?

a. CH_3-NH_3 b. HCN c. $NOBr$

2. What hybrid orbital set is used by the underlined atom in each of the following:

a. $\underline{Te}Cl_4$

b. $\underline{O}F_2$

c. $H\underline{C}N$

d. $\underline{I}Cl_2^+$

e. $\underline{I}F_5$

f. $\underline{Sn}Cl_2$

g. $\underline{I}F_4^+$

h. $\underline{Be}Cl_2$

i. $\underline{S}Cl_2$

j. $\underline{Si}F_6^{2-}$

k. $\underline{Sn}Cl_5^-$

l. $O\underline{C}S$

3.

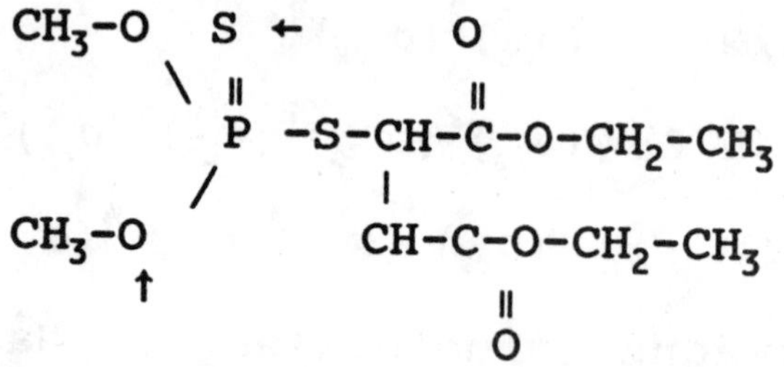

$$CH_3-O \quad S \leftarrow \quad O$$

$$\begin{array}{c} CH_3-O \quad \quad \quad \quad \\ \backslash \quad \| \quad \quad \quad \| \\ P -S-CH-C-O-CH_2-CH_3 \\ / \quad \quad \quad | \\ CH_3-O \quad \quad CH-C-O-CH_2-CH_3 \\ \uparrow \quad \quad \quad \quad \| \\ \quad \quad \quad \quad \quad \quad O \end{array}$$

Answer the questions below based on the hybridization of malathion, a commonly used pesticide, as shown above.

a. How many sigma bonds are present in this molecule? _____

b. How many pi bonds are present in this molecule? _____

c. What is the hybridization on the oxygen marked with an arrow? _____

d. What is the hybridization on the sulfur marked with an arrow? _____

4. Predict the type of hybridization and the carbon-carbon bond angle for each carbon in the following molecule:

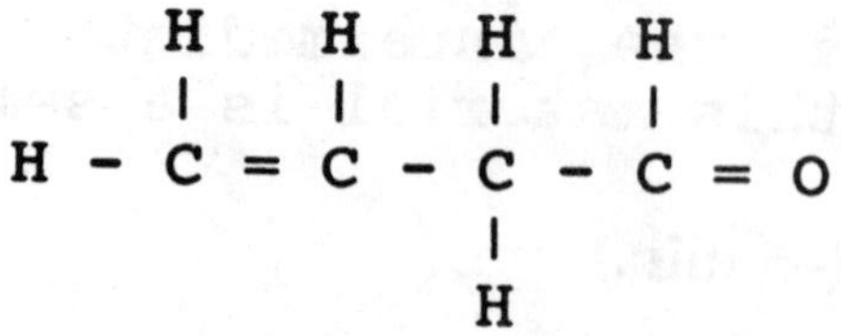

$$\begin{array}{c} H \quad H \quad H \quad H \\ | \quad | \quad | \quad | \\ H - C = C - C - C = O \\ | \\ H \end{array}$$

5. Based on the bond order predicted by Molecular Orbital Theory, arrange the following species in order of (a) increasing bond length and (b)

increasing dissociation energy: C_2^- C_2^+, and C_2.

6. Thionyl tetrafluoride, OSF_4, is the first sulfur compound for which structure determinations have indicated that five groups are bonded to a central sulfur atom. (a) Predict the molecular structure and the hybridization on the sulfur atom for this molecule. (b) Do you think it would be possible to synthesize a similar compound in which the sulfur were replaced by an oxygen atom? Do you think it would be possible to replace the sulfur with a selenium atom? Explain your answer in each case.

7. Determine the bond order and molecular orbital electron configuration for each of the following diatomic molecules or ions:

	Bond Order	Molecular Orbital Electron Configuration
a. CF	_____	_____
b. Cl_2^-	_____	_____
c. C_2^+	_____	_____
d. BeF	_____	_____
e. NO^+	_____	_____
f. B_2^+	_____	_____

8. Which of the molecules and molecule-ions in the previous question will be paramagnetic?

9. Identify each of the following elements as a conductor, an insulator, or a semiconductor.

a. silicon
c. potassium

b. sulfur
d. chromium

10. Cyclohexane, C_6H_{12}, like benzene, is an organic compound which has a molecular structure based on a six-membered ring of carbon atoms. Draw the structure of this molecule and determine the type of hybridization on the carbon atoms. The

hybridization on the carbon atoms in a benzene molecule is sp^2, and the resulting molecule is planar. Do you think that the cyclohexane molecule will be planar?

11. a. The energy separation between the valence band and the conduction band is called the _____. b. Substances that have a completely filled valence band are called _____. c. Semiconductor materials that are created by adding small amounts of other elements (called dopants) are known as _____ semiconductors. d. If the dopant has fewer valence electrons than the original substance, the resulting semiconductor is said to be a _____ semiconductor.

COMMENTS ON ARMCHAIR EXERCISES

1. The molecules discussed with the valence bond theory are generally more complicated than those discussed with the molecular orbital theory. This reflects the fact that the latter theory is more powerful but also more complex.

2. There are two possible ways that you might answer this question. You might focus on the relative size of the two central atoms, nitrogen and phosphorous, or you might consider the availability of the orbitals that would be necessary for the hybridization.

3. If the oxygen forms linear bonds, the hybridization must be sp. What might have happened to the p orbitals on the oxygen intead of forming the sp^3 hybridization that might be expected?

4. The molecular orbital description of the C_2 molecule would be

$$(\sigma_{2s})^2(\sigma^*_{2s})^2(\pi_{2p})^4$$

and the bond order predicts that this molecule would, indeed, be stable. If the emission of the blue light is caused by electrons dropping to this configuration from a more excited state, can you suggest where the electrons might have come from?

5. The almuminum works because it has one less valence electron than the silicon. Any other element with a similar electronic configuration, that is, in the same group of the periodic table, shuld act in a similar fashion and so could be subsituted for aluminum, i.e. boron.

CONCEPT TEST ANSWERS

1. sigma, pi
2. the number of pure atomic orbitals used in the combination.
3.

sp	linear
sp^3	tetrahedral
sp^3d^2	octahedral

4. an nonhybridized p atomic orbital
5. stereoisomers
6. delocalized
7.

hybrid type	no. of s orbitals	no. of p orbitals	no. of d orbitals	total hybrids
sp^2	1	2	0	3
sp^3d	1	3	1	5

8. the number of atomic orbitals brought by the combining atoms.
9. Bonding, antibonding
10. energy
11. homonuclear
12. antibonding, bonding
13. band
14. valence
15. semiconductors
16. Fermi level

PRACTICE TEST ANSWERS

1. **(L1)** a. sp^3 b. sp   c. sp^2

2. **(L1)**

a. $\underline{Te}Cl_4$ — dsp^3 b. $\underline{O}F_2$ — sp^3

c. $H\underline{C}N$ — sp   d. $\underline{I}Cl_2^+$ — sp^3

e. $\underline{I}F_5$ — d^2sp^3 f. $\underline{Sn}Cl_2$ — sp^2

g. $\underline{I}F_4^+$ — dsp^3 h. $\underline{Be}Cl_2$ — sp

i. $\underline{S}Cl_2$ — sp^3 j. $\underline{Si}F_6^{2-}$ — d^2sp^3

k. $\underline{Sn}Cl_5^-$ — dsp^3 l. $O\underline{C}S$ — sp

3. **(L1)**

$$CH_3-O \quad S \leftarrow sp^2 \quad O$$
$$\backslash \quad \parallel \qquad\qquad \parallel$$
$$P\ -S-CH-C-O-CH_2-CH_3$$
$$/ \qquad\qquad |$$
$$CH_3-O \qquad\qquad CH-C-O-CH_2-CH_3$$
$$\uparrow \qquad\qquad\qquad \parallel$$
$$\mathbf{sp^3} \qquad\qquad\qquad O$$

The molecule has 36 sigma bonds and 3 pi bonds.

4. **(L1)** Reading from left to right, the carbon hybridization is sp^2, sp^2, sp^3, and sp^2. (Notice that two carbon-carbon single bonds separate the double bonds, so there will not be significant delocalization.) The approximate bond angles (in the same order) will be $120°$, $120°$, $109.5°$, and $109.5°$.

5. **(L2)** In order of increasing bond length (bond order given in parentheses after each species):

$$C_2^- \ (2\ 1/2) < C_2 \ (2) < C_2^+ \ (1\ 1/2)$$

In order of increasing dissociation energy (bond order in parentheses):

$$C_2^+ \ (1\ 1/2) < C_2 \ (2) < C_2^- \ (2\ 1/2)$$

192

6. (L1) a. The hybridization on the sulfur atom is dsp^3 and the structure is a trigonal bipyramid.
b. Oxygen is unlikely to form a compound of this type, since it can't form an expanded valence shell. Selenium can form an expanded valence shell, and so the selenium compound is possible.

7. (L2)

	Bond Order	Molecular Orbital Electron Configuration
a. CF	2 1/2	$(\sigma_{2s})^2(\sigma^*_{2s})^2(\pi_{2p})^4(\sigma_{2p})^2(\pi^*_{2p})^1$
b. Cl_2^-	1/2	$(\sigma_{3s})^2(\sigma^*_{3s})^2(\pi_{3p})^4(\sigma_{3p})^2(\pi^*_{3p})^4(\sigma^*_{3p})^1$
c. C_2^+	1 1/2	$(\sigma_{2s})^2(\sigma^*_{2s})^2(\pi_{2p})^3$
d. BeF	2 1/2	$(\sigma_{2s})^2(\sigma^*_{2s})^2(\pi_{2p})^4(\sigma_{2p})^1$
e. NO^+	3	$(\sigma_{2s})^2(\sigma^*_{2s})^2(\pi_{2p})^4(\sigma_{2p})^2$
f. B_2^+	1/2	$(\sigma_{2s})^2(\sigma^*_{2s})^2(\pi_{2p})^1$

8. (L2) All of these are paramagnetic except NO^+.

9. (L3) Based on their positions on the periodic table, silicon is a semiconductor, sulfur is an insulator, and both potassium and chromium are conductors.

10. (L1)

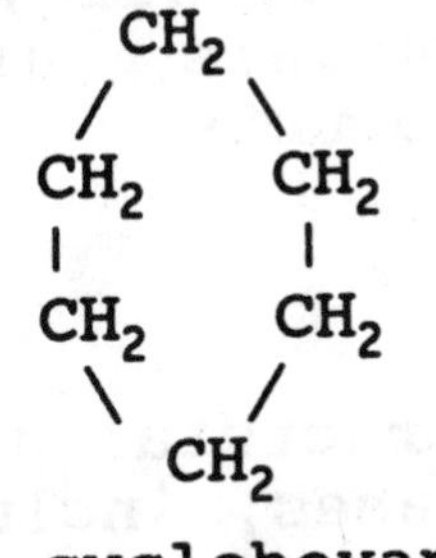

```
              CH2
             /    \
          CH2      CH2
           |        |
          CH2      CH2
             \    /
              CH2
```

The structure of cyclohexane is shown above. Each carbon atom displays sp^3 hybridization. Although the molecule appears to be planar, it is impossible to draw all of the carbon atoms in the same plane and maintain the 109.5° angles required by the sp^3 hybridization.

11. (L3) a. band gap   b. insulators,
        c. extrinsic   d. p-type

CHAPTER 11
BONDING AND MOLECULAR
STRUCTURE: ORGANIC CHEMISTRY

CHAPTER OVERVIEW

This chapter provides a relatively brief introduction to the field of organic chemistry. As you will see, the material is organized in a different way from most of the other topics in this book. Different classes of organic compounds react differently, and so the first step in organizing organic reactions is to be able to recognize these different classes of compounds.

A major focus of the discussion is the special nomenclature rules for each of these classes of compounds. Notice that just as each class of compound has slightly different nomenclature rules, each class also undergoes different chemical reactions. Therefore, being able to classify the classes of organic compounds is essential.

As you read this chapter, you will probably be impressed with how often organic chemistry is the basis for everyday life. Everything from insect stings to soap is explained by organic chemistry. In addition, synthetic organic polymers are a standard feature of everyday life, in the form of clothing, building materials, insulation, food wraps, and many other products.

LEARNING GOALS

1. Know how to write structural formulas for and also name the simple alkanes, including structural isomers. (Sec. 11.1)

2. Be able to write names for the common alkenes and alkynes, in addition to knowing some of the common reactions and preparation methods for these compounds. You should also recognize when stereoisomers are possible. (Sec. 11.2)

3. Be familiar with some of the nomenclature and reactions of aromatic compounds. (Sec. 11.3)

4. Be familiar with the nomenclature, methods of synthesis, and some common reactions of alcohols and ethers. (Sec. 11.4)

5. Be able to write names for carbonyl compounds, such as aldehydes, ketones, carboxylic acids, and esters, in addition to knowing some of the common reactions and preparation methods for these compounds. (Sec. 11.5)

6. You should be generally familiar with the types of fats and oils that are found in the foods that we eat, and also understand how soap is produced and why soap can contribute to the removal of grease and dirt. (Sec. 11.6)

7. Be familiar with the nomenclature, methods of synthesis, and common reactions of amines and amides. (Sec. 11.7)

8. Understand some of the ways in which synthetic organic polymers are used in our modern society, how they are classified, and some of the common methods by which they are produced and characterized. (Sec. 11.8)

ARMCHAIR EXERCISES

1. Many organic reactions will occur reasonably rapidly at modest temperatures (below 100 degrees Celsius). Aside from making it convenient to do organic reactions in the laboratory, does this have any other important results that you can think of?
2. Bromine dissolved in an organic solvent is a deep reddish color. Suppose that you have two identical test tubes of bromine solution, and add pentene to one and pentane to the other. Would you expect to see any differences?

3. In this chapter you have learned something about the type of compounds called polymers. You should recognize that in a polymer the molecular chains may be arranged in different ways. For some compounds the molecular chains are relatively straight, something like pieces of spaghetti before it is cooked. Other compounds have the chains intertwined, like well-cooked spaghetti that has

been thoroughly mixed up on the plate. As you might expect this produces different properties.

There is a simple way to observe this difference. Paper consists mainly of a natural polymer called cellulose. Carefully push a sharp pencil through a paper cup, from one side to the other, being careful hit the cup and not your own finger. Once the pencil has penetrated from one side to the other, pour water into the cup (OVER A SINK!). Does water leak around the pencil?

Next, repeat the process, but this time use a plastic zipper-lock bag, like those used to store food in a freezer. (The bag is probably made of a polymer called polyethylene.) Once you have carefully pushed the pencil through the bag from side to side, fill it with water and check for leaks. Is there any difference between the two containers? Does this suggest any difference between the two polymers?

CONCEPT TEST

1. Circle the unsaturated hydrocarbon.

A. $CH_3CH_2CH_3$ B. CH_4 C. CH_3CH_3 D. $CH_2=CH_2$

2. The hydrocarbon _____ is the principal constituent of natural gas.

3. Hydrocarbons with double and triple bonds between the carbon atoms are often referred to as

_____

4. _____ is the name of the alcohol found in alcoholic beverages.

5. Name the compounds having the structural formulas given.

A. $CH_3CH_2CH_3$ B. $CH_3CH_2CH_2CH_3$

C. $CH_3CH_2CH_2CH_2CH_2CH_3$ D. CH_3CH_3

6. The systematic names of alkanes are based on the number of carbon atoms in the _____ carbon chain.

7. _____ is the systematic name of the simple alkene having the formula $CH_2 = CH_2$.

8. The text discusses two different kinds of isomers. Compounds having the same formula but with the atoms are connected in a different order are called _____. Compounds that have the same formula and atom-to-atom connections, but have the atoms arranged differently in space are called _____.

9. When two substituents are attached to adjacent carbons on a benzene ring, they are said to be _____ ; when two substituents are attached to carbons that are separated by two carbon atoms on the benzene ring, they are said to be _____.

10. Ethylene and propylene are prepared industrially by heating the hydrocarbons found in natural gas and petroleum, a process called _____.

11. Benzene, toluene, and naphthalene are examples of the type of organic compounds that are called _____ because they usually have a pleasant odor.

12. When one or more of the hydrogens in a hydrocarbon is replaced with a(n) _____ group, the compound is called an alcohol.

13. _____ is the name of the compound commonly known as automobile antifreeze.

14. Aldehydes, ketones, and carboxylic acids all have a double-bonded carbon-oxygen grouping called a(n) _____ group.

15. _____ is the name of the most

important organic acid.

16. The odor of many fruits, such as bananas and pineapples, is produced by individual compounds in the general class of called _____.

17. The hydrolysis of esters is called a(n) _____ reaction, because in some cases the product is a soap.

18. _____ are the class of compounds responsible for the odor of decaying fish or decaying flesh.

19. The small molecules used to synthesize polymers are called _____.

20. Polymers that are obtained by polymerizing a mixture of two or more different monomers are called _____.

21. Match the organic functional groups in the list on the right below with the appropriate suffix or ending from organic nomenclature that corresponds to that functional group.

_____ -al   A. ketone

_____ -ene   B. aldehyde

_____ -yne   C. carbon-carbon double bond

_____ -oic acid D. alcohol

_____ -ol   E. organic acid

_____ -one   F. carbon-carbon triple bond

22. The refinery process called _____ is used to break down large alkanes into smaller, branched-chain alkanes; the process called _____ converts alkanes to aromatic compounds.

STUDY HINTS

1. Many problems in organic chemistry may require visualization of molecules in three dimensions, a skill best learned by working with molecular models. Take advantage of every opportunity to use the set in class or borrow models from a friend.

2. The repeating unit in a polymer is not just the simplest unit that is repeated but rather is determined by the monomer (or monomers) linked to create the polymer structure. For example, the repeating unit in polyethylene, shown below in brackets, is not CH_2 but rather CH_2CH_2, since that is the structure of the ethylene monomer.

$$- CH_2 - CH_2 \left[- CH_2 - CH_2 - \right] CH_2 -$$

3. Sometimes it's difficult to distinguish between addition and condensation polymerization based only on the formula of the repeating unit. Compare the polymerization reactions of ethylene glycol and ethylene oxide shown below.

ethylene glycol
$$HO - CH_2 - CH_2 - OH \rightarrow \left[- O - CH_2 - CH_2 - O - \right]$$
$$(-H_2O)$$

ethylene oxide
$$\overset{O}{\overset{\diagup \diagdown}{CH_2 - CH_2}} \rightarrow \left[- O - CH_2 - CH_2 - O - \right]$$

The product appears to be the same in both cases, but the first reaction is a condensation mechanism and the second is an additive mechanism. (Notice, however, that polymers having the same repeating unit may have different properties if they have been formed by different reactions.)

PRACTICE PROBLEMS

1. (L5) Which of the following compounds can form an acid when it is oxidized?

a. 2-pentanol b. butanone c. hexanal d. 3-pentanol

2. (L1, L2, L3, L4, L5, and L7) Write systematic names for each of the following compounds.

a. CH_3-O-CH_3 _____

b. $CH_3CH=CHCH_3$ _____

c. CH_3CH-OH _____
     |
     CH_3

d.   O
             ‖
$CH_3CH_2CH_2CCH_3$ _____

e. NH_2

_____

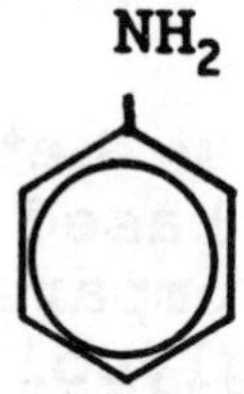

f. CH_3

          CH_3 _____

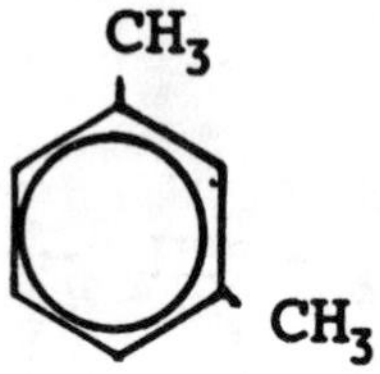

3. (L1) Draw all of the structural isomers having the formula C_6H_{14}.

4. (L1) Name each of the structural isomers you drew in the previous question.

5. (L1, L3, L4, L5, and L7) Draw the structure of each of the following compounds:

a. 1-pentyne

b. 1,2-diaminoethane

c. oxalic acid

d. 2-hexanol

e. propanoic acid

f. ethyl acetate

g. 2-butanone

h. o-dichlorobenzene

6. (L1 thru L7) Write the correct product(s) to complete the following equations.

a. CH_3CH_2OH + H_2SO_4 →

b. $CH_3CH_2\overset{\displaystyle O}{\overset{\displaystyle \|}{C}}H$ $\xrightarrow{K_2Cr_2O_7}$

c. $CH_3CH_2\overset{\displaystyle OH}{\overset{\displaystyle |}{C}}HCH_3$ $\xrightarrow{KMnO_4}$

d. $CH_3CH=CHCH_2CH_3$ + $H_2(g)$ $\xrightarrow{\text{metal catalyst}}$

e. $CH_2=CHCH_2CH_3$ + HCl →

7. **(L8)** Identify each of these reactions as either an addition polymerization or a condensation polymerization reaction.

a. $CF_2 = CF_2$ → $[- CF_2 - CF_2 -]$

b. $CH_2 = CH - CH = CH_2$ → $[- CH_2 - CH = CH - CH_2 -]$

c. $HO-CH_2 - \bigcirc - CO_2H$ → $[- O-CH_2 - \bigcirc - \overset{O}{\underset{||}{C}} -]$

8. **(L8)** Place a bracket around the repeating unit for each of the following simple addition polymers.

a.
$$-CH_2-\underset{\underset{CH_3}{|}}{\overset{\overset{CH_3}{|}}{C}}-CH_2-\underset{\underset{CH_3}{|}}{\overset{\overset{CH_3}{|}}{C}}-CH_2-\underset{\underset{CH_3}{|}}{\overset{\overset{CH_3}{|}}{C}}-$$

b.
$$-CH_2-\underset{}{\overset{\overset{CN}{|}}{CH}}-CH_2-\underset{}{\overset{\overset{CN}{|}}{CH}}-CH_2-\underset{}{\overset{\overset{CN}{|}}{CH}}-$$

c.
$$-CH_2-\underset{\underset{Cl}{|}}{\overset{\overset{Cl}{|}}{C}}-CH_2-\underset{\underset{Cl}{|}}{\overset{\overset{Cl}{|}}{C}}-CH_2-\underset{\underset{Cl}{|}}{\overset{\overset{Cl}{|}}{C}}-CH_2-\underset{\underset{Cl}{|}}{\overset{\overset{Cl}{|}}{C}}-$$

d.
$$-\underset{}{\overset{\overset{CH_3}{|}}{CH}}-O-\underset{}{\overset{\overset{CH_3}{|}}{CH}}-O-\underset{}{\overset{\overset{CH_3}{|}}{CH}}-O-\underset{}{\overset{\overset{CH_3}{|}}{CH}}-O$$

PRACTICE PROBLEM SOLUTIONS

1. c. hexanal

2. a. dimethylether
   b. 2-butene (note cis and trans isomers are possible)
   c. 2-propanol
   d. 2-pentanone
   e. aniline
   f. **m**-dimethylbenzene

3. and 4.
$CH_3CH_2CH_2CH_2CH_2CH_3$

n-hexane

$CH_3CHCH_2CH_2CH_3$
$\quad\quad |$
$\quad\quad CH_3$

2-methylpentane

$CH_3CH_2CHCH_2CH_3$
|
CH_3

3-methylpentane

$CH_3CHCHCH_3$
| |
H_3C CH_3

2,3-dimethylbutane

CH_3
|
$CH_3CCH_2CH_3$
|
CH_3

2,2-dimethylbutane

5. a. $HC \equiv CCH_2CH_2CH_3$

b. $NH_2CH_2CH_2NH_2$

c. $HO-\overset{\overset{\displaystyle O}{\|}}{C}-\overset{\overset{\displaystyle O}{\|}}{C}-OH$

d. $CH_3CHCH_2CH_2CH_2CH_3$
|
OH

e. $CH_3CH_2\overset{\overset{\displaystyle O}{\|}}{C}-OH$

f. $CH_3\overset{\overset{\displaystyle O}{\|}}{C}-O-CH_2CH_3$

g. $CH_3CH_2\overset{\overset{\displaystyle O}{\|}}{C}CH_3$

h. 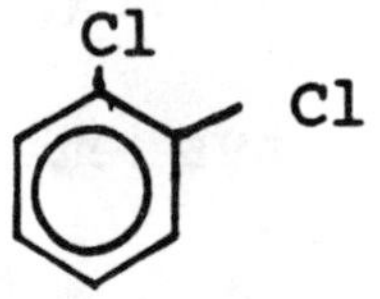

6. a. $CH_3CH_2OH + H_2SO_4 \rightarrow CH_2{=}CH_2 + HOH$

(depending upon conditions, the product may be $CH_3CH_2OCH_2CH_3$)

b. $CH_3CH_2\overset{\overset{\displaystyle O}{\|}}{C}H \xrightarrow{Na_2Cr_2O_7} CH_3CH_2\overset{\overset{\displaystyle O}{\|}}{C}-OH$

c. $CH_3CH_2\overset{\overset{\displaystyle OH}{|}}{C}HCH_3 \xrightarrow{KMnO_4} CH_3CH_2\overset{\overset{\displaystyle O}{\|}}{C}CH_3$

d. $CH_3CH=CHCH_2CH_3 + H_2(g)$ $\xrightarrow{\text{metal catalyst}}$ $CH_3CH_2CH_2CH_2CH_3$

e. $CH_2=CHCH_2CH_3 + HCl \rightarrow CH_3CHCH_2CH_3$
$$\underset{Cl}{\overset{|}{}}$$

8. Reactions a and b are simple addition, since there is no removal of a water molecule. In reaction c, however, a water molecule is removed, so this is a condensation reaction.

9. a.

$$-CH_2-\underset{\underset{CH_3}{|}}{\overset{\overset{CH_3}{|}}{C}}\left[-CH_2-\underset{\underset{CH_3}{|}}{\overset{\overset{CH_3}{|}}{C}}-\right]CH_2-\underset{\underset{CH_3}{|}}{\overset{\overset{CH_3}{|}}{C}}-$$

b.

$$-CH_2\left[\overset{\overset{CN}{|}}{CH}-CH_2\right]\overset{\overset{CN}{|}}{-CH}-CH_2-\overset{\overset{CN}{|}}{CH}-$$

c.

$$-CH_2-\underset{\underset{Cl}{|}}{\overset{\overset{Cl}{|}}{C}}\left[CH_2-\underset{\underset{Cl}{|}}{\overset{\overset{Cl}{|}}{C}}-\right]CH_2-\underset{\underset{Cl}{|}}{\overset{\overset{Cl}{|}}{C}}-CH_2-\underset{\underset{Cl}{|}}{\overset{\overset{Cl}{|}}{C}}-$$

d.

$$-\overset{\overset{CH_3}{|}}{CH}-O\left[-\overset{\overset{CH_3}{|}}{CH}-O-\right]\overset{\overset{CH_3}{|}}{CH}-O-$$

PRACTICE TEST (30 Minutes)

1. Write systematic names for each of the following compounds.

a. $CH_3CH_2-O-CH_2CH_3$ _____

b. CH_3CHCH_3 _____
$$\underset{CH_3}{\overset{|}{}}$$

c. $CH_2=CHCH=CH_2$ _____

d. $CH_3C \equiv CCH_3$ _____

e. CH_2Cl_2 _____

f. CH_3CH_2COOH _____

204

2. Draw the structure of each of the following compounds:

a. 3-hexene
c. 1,3-dibromobutane
e. 2-pentanol
g. ethanoic acid

b. heptanal
d. 2-butanone
f. ethyl butyrate
h. methylamine

3. (L7) Based on your knowledge of Valence Bond Theory, answer the questions below regarding the structure of the pesticide Baygon, which is shown on the right.

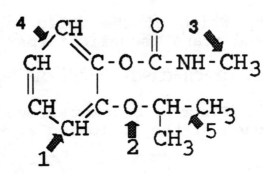

a. Describe the type of hybridization of the carbon atom labeled #1.

_____

b. Describe the type of hybridization on the oxygen atom labeled #2. _____

c. Describe the type of hybridization on the carbon atom labeled #3. _____

d. Which carbon-carbon bond is shorter, the one labeled #4, the one labeled #5, or are they the same length? _____

4. (L1 thru L7) Write the correct product(s) to complete the following equations.

a. $CH_3CH_2\overset{\overset{\displaystyle O}{\|}}{C}CH_3$ $\xrightarrow{\text{LiAlH}_4}$

b. $H_2C=CH_2$ + Br_2 →

c. $CH_3CH_2CH_2-OH$ $\xrightarrow{H_2SO_4/180°}$

d. $CaC_2(s) + 2 H_2O(\ell)$ →

e. $CH_3CH-CHCH_2CH_3$ + KOH(ethanol) $\rightarrow$
        | |
        H   Cl

f. CH_3NH_2 + $CH_3CH_2\overset{\overset{O}{\|}}{C}-OH$ $\rightarrow$

5. Which of the following compounds can have cis and trans isomeric forms?

a. $Cl-CH=CH-Cl$ b. $Cl-CH_2-CH_2-Cl$ c. $Cl-C \equiv C-Cl$

6. Place a bracket around the repeating unit for each of the following simple addition polymers.

a.

$-CH_2-O-CH_2-O-CH_2-O-$

c.
```
   F F F F F F F
   | | | | | | |
  -C-C-C-C-C-C-C-
   | | | | | | |
   F F F F F F
```

d.
```
      CH3      CH3      CH3
      |        |        |
-CH2 -CH-CH2 -CH-CH2 -CH-
```

COMMENTS ON ARMCHAIR EXERCISES

1. Organic chemistry is sometimes called the chemistry of living things (although it is now much broader than that).   The fact that organic reactions will normally occur with reasonable speed at moderate temperatures makes possible the wealth of organic reactions which are the basis of life.

2. The pentene will react readily with the bromine causing that solution to lose its color rather rapidly. The pentane solution will change color much more slowly.

3. If you have done this carefully, you should have seen a drip of water around the pencil in the paper cup but very little leakage around the pencil in the plastic bag. The polyethylene molecules are tangled together, so that the sharp point of the

206

pencil pushes the molecules aside rather than breaking them. (This experiment is based on a description in *A+ Projects for Chemistry* by Janice VanCleave. This book also contains many other projects that may be interesting to you.)

Magicians sometimes use a special sharp needle coated with a lubricant to penetrate an inflated balloon without breaking it or allowing the air to escape. What does this suggest about the structure of the chains in the polymer that is used to make the balloon?

CONCEPT TEST ANSWERS

1. D. $CH_2=CH_2$
2. methane
3. unsaturated
4. ethanol
5. A. n-Propane B. n-Butane
   C. n-Hexane D. Ethane
6. longest continuous
7. ethene
8. structural isomers, stereoisomers
9. ortho, para
10. steam cracking
11. aromatics
12. -OH or hydroxyl
13. ethylene glycol
14. carbonyl
15. acetic acid
16. esters
17. saponification
18. amines
19. monomers
20. co-polymers
21. -al   B. aldehyde
    -ene   C. carbon-carbon double bond
    -yne   F. carbon-carbon triple bond
    -oic acid E. organic acid
    -ol   D. alcohol
    -one   A. ketone
22. catalytic cracking, catalytic reforming

PRACTICE TEST ANSWERS

1. a. diethyl ether
   b. 2-methylpropane (isobutane)
   c. 1,3-butadiene   d. 2-propyne
   e. dichloromethane   f. propanoic acid

2. a. $CH_3CH_2CH=CHCH_2CH_3$ b. $CH_3CH_2CH_2CH_2CH_2CH_2\overset{\displaystyle O}{\overset{\displaystyle \|}{C}}H$

c. $BrCH_2CH_2CHCH_3$
               |
               Br

d.   O
           ||
     $CH_3CCH_2CH_3$

e. $CH_3CHCH_2CH_2CH_3$
      |
      OH

f.   O
          ||
    $CH_3CH_2CH_2COCH_2CH_3$

g.   O
     ||
  CH_3C-OH

h. CH_3NH_2

3. a. sp^2 b. sp^3 c. sp^3

d. bond #4 is longer; the carbon-carbon bond in the benzene ring is not really a single bond, but has a bond order of about 1.5.

4.   O   OH
      || $LiAlH_4$ |
a. $CH_3CH_2CCH_3$ $\rightarrow$ $CH_3CH_2CHCH_3$

b. $H_2C=CH_2$ + Br_2 $\rightarrow$ $BrCH_2CH_2Br$

                       $H_2SO_4/180°$
c. $CH_3CH_2CH_2-OH$ $\rightarrow$ $CH_3CH=CH_2$ + H_2O

d. $CaC_2(s) + 2\ H_2O(\ell) \rightarrow H-C \equiv C-H(g)$ + $Ca(OH)_2(s)$

e. $CH_3CH-CHCH_2CH_3$ +KOH(ethanol) $\rightarrow$ $CH_3CH=CHCH_2CH_3$
       | |
      H Cl

                     O   O
                    || ||
f. $CH_3NH_2 + CH_3CH_2C-OH \rightarrow$ $CH_3CH_2C-NH-CH_3$ + HOH

5. A. Cl-CH=CH-Cl, There would be free rotation around the carbon-carbon bond in the second compound. There is no rotation around the triple bond, and the molecule is linear.

6. (L8)
a.

b.

c.

$$
\begin{array}{ccccccc}
\text{F} & \text{F} & \text{F} & \text{F} & \text{F} & \text{F} & \text{F} \\
| & | & | & | & | & | & | \\
-\text{C}-\text{C}\!\!-\!\!\left[\text{C}-\text{C}\right]\!\!-\!\!\text{C}-\text{C}-\text{C}- \\
| & | & | & | & | & | & | \\
\text{F} & \text{F} & \text{F} & \text{F} & \text{F} & \text{F} & \text{F}
\end{array}
$$

d.

$$
\begin{array}{ccc}
 & \text{CH}_3 & \text{CH}_3 & \text{CH}_3 \\
 & | & | & | \\
-\text{CH}_2\!\!-\!\!\left[\text{CH}-\text{CH}_2\right]\!\!-\!\!\text{CH}-\text{CH}_2-\text{CH}-
\end{array}
$$

CHAPTER 12
GASES

CHAPTER OVERVIEW

Gases are largely invisible but play a crucial role in our everyday life, from the air that we breath to the rust that forms on our automobiles. Gases were mathematically described before solids or liquids and are still the best understood of the three states of matter.

The quantitative study of gases began in the 17th century when Robert Boyle described the relationship between pressure and volume, but it was over a hundred years before Charles recognized the comparable relationship between temperature and volume. The long delay was due to the absence of a temperature scale, like kelvins, that would show how temperature and volume were related. Even today, students run afoul of this same problem when they forget to convert temperatures to kelvins.

The mole relationship, which has been used frequently in the earlier parts of the book, was first recognized by Avogadro, as he was trying to understand the behavior of gases. Thus, Avogadro's law is already familiar to you, although in a different form. The combination of the individual gas laws produces the ideal gas law, which is a key description of gas behavior. This also leads to the recognition that volumes can be used in stoichiometric calculations in the same way that moles were used earlier, as long as the substances involved are gases, and temperature and pressure are held constant.

The kinetic molecular theory is important not only for an understanding of gases, but also as a basis for topics that will be discussed later. It's easy to forget that, even though the kinetic theory seems like common sense to us today, this theory was once considered to be controversial. The acceptance of the basic ideas of this theory was an important step on the road towards the atomic theory that we discussed earlier. The diffusion or effusion of gases is important evidence that there is, indeed, continuous molecular motion in a gas.

It may be surprising to you that, despite all of the time spent learning the various gas laws, most of them don't work exactly with real gases. That's why we call the summary law the ideal gas law. Under normal conditions the errors are relatively small, but if the pressure is very high or the temperature is very low, the errors become very significant. One of the strengths of the kinetic molecular theory is that it helps to explain why these problems exist for real gases.

LEARNING GOALS

1. Understand how gas pressure is measured using a barometer or a manometer, and recognize the relationship among the various units of pressure, including millimeters of mercury, torr, atmospheres, pascals, and kilopascals. If necessary, review temperature conversions to insure that there will be no difficulty doing conversions involving degrees Celsius, degrees Fahrenheit, and kelvins. (Sec. 12.1)

2. Understand the simple gas laws, including Boyle's law, Charles's law, and Avogadro's law. (Sec. 12.2)

3. Avogadro's law allows us to relate gas volumes to moles of gas and number of gas molecules. These relationships are often helpful when working problems. (Sec. 12.2)

4. The three gas laws mentioned in learning goal 2 can be combined to produce a useful relationship called the ideal gas law. This relationship can be used to solve many different types of gas law problems. Be especially sure to understand problems relating to gas densities or molar masses of gases, since the ideal gas law is particularly important for these calculations. (Sec. 12.3)

5. Another possible combination of the simple gas laws is called the general gas law. Compare this equation carefully with the ideal gas law to see when each will be most useful. (Sec. 12.3)

6. Stoichiometric calculations for chemical equations involving gases can be worked by using gas volumes instead of moles. It's important to be able to use both calculation methods. (Sec. 12.4)

7. Understand how to do gas law calculations for mixtures of gases based on Dalton's law of partial pressures. (Sec. 12.5)

8. The kinetic molecular theory provides the basis for the currently accepted explanation of gas behavior. Understand the assumptions of this theory and also the important new terms related to the theory, such as average molecular kinetic energy, average molecular speed, and the distribution of molecular speeds. (Sec. 12.6)

9. Diffusion or effusion of gases is described in terms of Graham's law; know how to solve problems based on this relationship. (Sec. 12.7)

10. Understand why gas behavior doesn't follow the predictions of the ideal gas law and how to use the van der Waals equation to improve on these predictions. (12.9)

ARMCHAIR EXERCISES

1. In his delightful book, *Kitchen Science*, Howard Hillman asks the reader to explain why the lid may stick on a covered pot containing a little water which has been boiled for a while and then allowed to thoroughly cool. He notes that neither trying to pry off the lid nor putting the pot in cold water is likely to loosen the lid. Can you explain these observations.

2. Used 35 mm plastic film canisters (about two inches high) are usually available from a camera shop and provide an interesting demonstration of gas behavior. When the canister is filled about half full with water, then one-half of an Alka-Seltzer tm tablet is added and the cap put on, a small explosion will result in a short time. (IF YOU TRY THIS BE SURE NOT TO STAND TOO CLOSE AND DO THE REACTION WHERE THE LIQUID SPRAY WILL NOT CAUSE DAMAGE!) Why did the canister explode?

3. Suppose that you use an ice pick or a very large needle to place three holes in a two-Liter soda bottle, one about an inch from the bottom, one about midway up the side of the bottle, and one about two thirds of the way up the side. (IF YOU TRY THIS, BE SURE TO PUNCTURE THE BOTTLE AND NOT YOUR FINGER!) When the bottle is filled with water, three streams of water will come from these three holes, but the three streams don't behave the same way. The one from the top of the bottle doesn't squirt out as far as the one in the middle, and the bottom stream squirts out furthest of all. Can you explain why this is true?

Does this experience seem to bear any relationship to the fact that air pressure is less on a mountain top than it is in the valley?

4. The May 1990 issue of the US/Russian science journal, *Quantum*, includes several questions that can be answered using the kinetic molecular theory. For example, it points out that at high altitudes in the Earth's atmosphere, the molecules of air have velocities that are equivalent to a temperature above a thousand degrees Celsius. Why don't orbiting satellites melt as a result of being exposed to these high temperatures?

CONCEPT TEST

1. One standard atmosphere is equivalent to values of _____ mm of Hg or _____ kilopascals.

2. _____ is defined as the force exerted on an object divided by the area over which the force is exerted.

3. According to Boyle's law, if the pressure is doubled on a gas sample, with no change in the temperature or amount of gas, the volume will

_____.

4. Boyle's law states that the volume of a fixed amount of gas at a given temperature is inversely

proportional to the _____ exerted on the gas.

5. Charles's law states that if a given quantity of gas is held at constant pressure, its volume is directly proportional to the _____.

6. According to Charles's law, if the temperature is doubled on a gas sample, with no change in the pressure or amount of gas, the volume will

_____.

7. If the volume of a gas sample is graphed against the kelvin temperature while pressure is constant, a straight line results. By extrapolation to zero volume, this line results in a temperature value of _____ °C, which is called absolute zero.

8. Avogadro stated that equal volumes of gases under the same conditions of temperature and pressure have equal numbers of _____.

9. According to Avogadro's law, if the number of moles of gas is doubled, with no change in the pressure or temperature, the volume will

_____.

10. Under conditions of standard temperature and pressure (STP), 1.0000 mole of any gas occupies a volume of _____ liters, a quantity called the

_____.

11. The value of the gas constant, R, in the ideal gas equation is _____. (units?)

12. What does STP stand for? _____

13. If the temperature on a confined gas sample is doubled, while the volume is held constant, what will happen to the pressure? _____

214

14. At 25.0°C a sample of nitrogen gas contained in a 1.000 Liter container has a pressure of 345 mmHg, and also at 25.0°C a sample of hydrogen gas contained in a 1.000 Liter container has a pressure of 265 mmHg. If the two gas samples are combined in a 1.000 Liter container at the same temperature, what pressure will result? _____

15. Dalton's law of partial pressures states that the total pressure exerted by a mixture of gases is the sum of the _____ of the individual gases in the mixture.

16. When a gas is collected over water, the actual pressure of the gas collected is equal to the atmospheric pressure minus _____.

17. The observation that gases are compressible suggests that the distance between the gas particles (atoms or molecules) is _____ compared to the actual size of the particles.

18. For a given sample of gas molecules, the average kinetic energy depends only on the value of the _____.

19. When the temperature of a gas sample is increased, the average velocity of the gas molecules is _____.

20. The assumptions of the kinetic molecular theory are most likely to be incorrect for gases under which of the following combinations of conditions?
    a. high temperature, high pressure
    b. high temperature, low pressure
    c. low temperature, high pressure
    d. low temperature, low pressure

21. The values of the a and b constants in the van der Waals Equation are shown below for some typical gases.

gas	a	b
CCl_4	20.4 $L^2 \cdot atm/mol^2$	0.1383 L/mol
Kr	2.32	0.0398
O_2	1.36	0.0318

a. According to the values above, which of these three gases would you expect to display the smallest intermolecular attraction?   _____

b. Which gas has molecules that you would expect to occupy the largest volume?   _____

STUDY HINTS:

1. In order to work gas law problems, it is frequently necessary to organize many different numerical values. The best way to do this is by using a data table, and this procedure is followed throughout the textbook and this study guide. You may wish to use a modified version of the data table form shown here, but you will definitely want to use some type of data table to organize the information in these problems.

2. Neither Celsius nor Fahrenheit temperatures are useful for gas law calculations; *you must use kelvins*. Unfortunately it is easy to forget this when you become involved in a problem. To avoid this mistake, always convert to kelvins as soon as possible and insert these values in the data table.

3. Students seem to be most likely to forget to convert to kelvins if the temperature is 273°C. It's not clear why this is true, but watch out for the situation and avoid this error.

4. Students sometimes forget the correct units that are required when values are inserted into the ideal gas law. If you memorize not just the numerical value but also the units of the gas constant, R, this should never be a problem.

5. You have probably noticed that the value of standard temperature is different for thermodynamics problems (25°C) than for gas law problems (0°C). Be sure to think about which problem type you are working so that you won't be confused by this difference.

PRACTICE PROBLEMS

1. (L1) A certain gas sample is measured at 720.0 mmHg and 25°C. Calculate the corresponding temperature in kelvins and pressure in atmospheres.

2. (L2) A sample of oxygen gas has a volume of 128 milliliters at a pressure of 500.0 mmHg. Calculate the volume this same sample would occupy at standard pressure.

3. (L2) A sample of nitrogen gas has a volume of 212 milliliters at a temperature of 57.0°C. Calculate the volume this same sample would occupy at standard temperature.

4. (L4) The density of an unknown gas measured at standard temperature and pressure is 1.96 grams/liter. What is the molar mass of this gas?

5. (L3) A high volume sampler used for air pollution work analyzes 1200 cubic meters of air. If this volume was measured at STP, how many molecules of air were sampled?

6. (L5) A certain mass of chlorine gas occupies a volume of 50.0 milliLiters at a temperature of $273°C$ and a pressure of 700.0 mmHg. What volume will this same sample of gas occupy at STP?

7. (L4) How many moles of nitrogen gas are contained in a sample vial having a volume of 1.78 L if the temperature of the gas is $27.0°C$ and the pressure is 750.0 mmHg?

8. (L7) A mixture of gases consists of 56.0 grams of N_2, 16.0 grams of CH_4, methane, and 48.0 grams of O_2. If the total pressure of this mixture is 850.0 mmHg, what is the mole fraction and partial pressure of each gas.

9. (L4) Calculate the pressure (in kilopascals) which will result if 2.5 grams of XeF_4 gas is introduced into an evacuated container which has a volume of 3.00 cubic decimeters and is kept at a constant temperature of $80.0°C$. (In SI units,

$$R = 8.31 \frac{kPa \cdot dm^3}{mol \cdot K})$$

10. (L9) An unknown gas diffuses through a small opening at a rate of 23 mL/hour, but helium gas diffuses through the same opening at a rate of 92 mL/hour. What is the molar mass of the unknown gas?

11. (L6) The empirical formula of a certain hydrocarbon is CH_3, and when 0.500 mole of this hydrocarbon is completely combusted, 67.2 liters of carbon dioxide (measured at STP) is collected. What is the molecular formula of this hydrocarbon?

12. **(L7)** What is the pressure that results when 2.0 liters of hydrogen gas, measured at STP, is forced into a 2.0 liter container, which previously contained enough oxygen to completely fill the container under standard conditions. Assume that the temperature doesn't change when the gases are mixed.

13. **(L6)** At 383 K, methane, an organic compound, reacts with oxygen to form only gaseous products as indicated in the following equation:

$$CH_4(g) \ + \ O_2(g) \ \rightarrow \ CO(g) \ + \ H_2O(g)$$

If 0.010 mole of methane and 0.030 mole of oxygen are placed in a sealed container at an initial pressure of 1.00 atmospheres and the temperature is held constant during the reaction, calculate (a) the volume of the container, (b) the moles of carbon monoxide formed, and (c) the total pressure in the container at the end of the process.

PRACTICE PROBLEM ANSWERS

1. pressure = $\dfrac{720.0 \text{ mmHg}}{760.0 \text{ mmHg/atm}}$ = 0.947 atm

   temperature = 25°C + 273 = 298 kelvin

2. Remember that standard pressure is 760 mmHg and use Boyle's law.

   $$V = 128 \text{ mL} \times \frac{500.0 \text{ mmHg}}{760.0 \text{ mmHg}}$$

   V = 84.2 mL

3. Use Charles's law. Don't forget to convert the temperature to kelvins.

   $$V = 212 \text{ mL} \times \frac{273.2 \text{ K}}{(57.0 + 273.2) \text{ K}} = 175 \text{ mL}$$

4. First set up the data table

mass	1.96 g
volume	1.00 L
temp	273 K
pressure	1.00 atm
molar mass	?

   The ideal gas law contains all of these variables.

   $$PV = nRT = \frac{gRT}{M}$$

   rearrange the equation to solve for molar mass

   $$M = \frac{gRT}{PV}$$

   $$M = \frac{1.96 \text{ g} \times 0.0821 \text{ L·atm/mol·K} \times 273 \text{ K}}{1.00 \text{ atm} \times 1.00 \text{ L}}$$

   M = 43.9 g/mol

Notice that there is also an alternative method of solution, based on the observation that 1 mole of any gas at STP must occupy a volume of 22.414 L.

$$M = 1.96 \text{ g/L} \times 22.41 \text{ L/mol}$$

$$\underline{M = 43.9 \text{ g/mol}}$$

5. First you must convert 1200 m^3 into liters. Since a cubic meter is a cube 1 meter on a side, and one meter is 100 cm, the volume of a cubic meter in cubic centimeters is

$$V = 100 \text{ cm} \times 100 \text{ cm} \times 100 \text{ cm} = 1 \times 10^6 \text{ cm}^3$$

One cm^3 = one mL, so the volume is also 1×10^6 mL.

Thus, 1 m^3 is equivalent to 1000 L.

Using this conversion factor,

$$V = 1200 \text{ m}^3 \times 1000 \text{ L/m}^3 = 1.2 \times 10^6 \text{ L}$$

Since the volume is measured at STP, we can use the molar volume, 22.4 L/mol, to determine how many moles of air were sampled.

$$\text{moles} = \frac{1.2 \times 10^6 \text{ L}}{22.4 \text{ L/mol}} = 5.36 \times 10^4 \text{ mol}$$

Multiply the number of moles by Avogadro's number to obtain the number of molecules.

no. of molecules
$$= 5.36 \times 10^4 \text{ mol} \times 6.02 \times 10^{23} \text{ molecules/mol}$$

$$\underline{\text{no. of molecules} = 3.2 \times 10^{28} \text{ molecules}}$$

6. First, set up a data table.

	Initial conditions	Final conditions
Volume	50.0 mL	?
Pressure	700.0 mmHg	760 mmHg
Temperature	273 + 273 = 546 K	273 K
Moles	n_1	n_1

The data provided will fit well into the general gas law.

$$\frac{P_1V_1}{nT_1} = \frac{P_2V_2}{nT_2}$$

Now substitute the available data into the equation. Since the number of moles doesn't change, it will cancel out.

$$\frac{700.0 \text{ mmHg} \times 50.0 \text{ mL}}{546 \text{ K}} = \frac{760.0 \text{ mmHg} \times V}{273 \text{ K}}$$

Rearranging to isolate V, the unknown.

$$V = \frac{700.0 \text{ mmHg} \times 50.0 \text{ mL} \times 273 \text{ K}}{546 \text{ K} \times 760.0 \text{ mmHg}}$$

$$V = 23.0 \text{ mL}$$

7. Again it is best to start by writing a data table. This time we will use a slightly different format to show that there is more than one way to organize a data table.

Volume = 1.78 L   Temperature = 27 + 273 = 300 K

Pressure = $\frac{750 \text{ mmHg}}{760 \text{ mmHg/atm}}$ = 0.987 atm

Moles = ?

In this case, the data is appropriate for substitution into the ideal gas law, PV = nRT. Rearranging to isolate the unknown, n, produces

$$n = \frac{PV}{RT}$$

Substituting the values into this relationship,

$$n = \frac{0.987 \text{ atm} \times 1.78 \text{ L}}{0.08206 \text{ L·atm/mol·K} \times 300 \text{ K}}$$

$$n = 0.0713 \text{ moles}$$

8. Since the grams of each gas is given, the obvious way to solve is to calculate the moles of each gas, then add to obtain the total moles, and finally determine mole fractions. Notice the data table looks different, but it still serves to organize the available information.

$$\text{mol } N_2 = \frac{56.0 \text{ g}}{28.02 \text{ g/mol}} = 1.999 \text{ mol}$$

$$\text{mol } CH_4 = \frac{16.0 \text{ g}}{16.04 \text{ g/mol}} = 0.998 \text{ mole}$$

$$\text{mol } O_2 = \frac{48.0 \text{ g}}{32.0 \text{ g/mol}} = 1.50 \text{ mol}$$

Total moles = mol N_2 + mol O_2 + mol CH_4

Total moles = 1.999 mol + 1.50 mol + 0.998 mol

Total moles = 4.497 mol

Now find the mole fraction of each gas.

Mole fraction N_2 = 1.999 mol/4.497 mol = <u>0.445 mol</u>

Mole fraction CH_4 = 0.998 mol/4.497 mol = <u>0.222 mol</u>

Mole fraction O_2 = 1.50 mol/4.497 mol = <u>0.334 mol</u>

Notice that the sum of all of the mole fractions equals 1.0, as is required by the definition of mole fraction.

To find the partial pressure of each gas, multiply the mole fraction times the total gas pressure.

Partial pressure N_2 = 0.445 x 850 mmHg

= <u>378 mmHg</u>

Partial pressure CH_4 = 0.222 x 850 mmHg

= <u>189 mmHg</u>

Partial pressure O_2 = 0.334 x 850 mmHg

= __284 mmHg__

9. As before, it is best to begin with a data table.

Temperature = 80.0 + 273.2 = 353.2 K
Pressure   = ?
Volume   = 3.00 dm^3

For XeF_4, the molar mass is 207.3 g/mol

Thus,
moles of XeF_4 = 2.50 g / 207.3 g/mol = 0.0121 moles

All of the data needed for the ideal gas law is given, except that the volume is in SI units. Since the pressure is requested in kPa, and the appropriate value for R is also given, the problem can be solved by using the ideal gas law with SI units.

Rearranging PV = nRT to solve for P,

$$P = \frac{nRT}{V}$$

Substituting the values into this relationship,

$$P = \frac{0.0121 \text{ mol} \times 8.31 \text{ kPa} \cdot \text{dm}^3/\text{mol} \cdot \text{K} \times 353 \text{ K}}{3.00 \text{ dm}^3}$$

__P = 12 kPa__

10. Problems that concern diffusion or effusion are almost always based on Graham's law. To avoid confusion, be sure to set up a data table that defines which gas will be labelled 1 and which will be labelled 2.

	Molar mass	Rate
1. unknown	?	23 mL/hr
2. helium	4.00 g/mol	92 mL/hr

225

Graham's law may be stated

$$\frac{\text{rate}_1}{\text{rate}_2} = \frac{\sqrt{M_2}}{\sqrt{M_1}}$$

inserting the values

$$\frac{23 \text{ mL/hr}}{92 \text{ mL/hr}} = \frac{\sqrt{4.00 \text{ g/mol}}}{\sqrt{M_1}}$$

$$M_1 = (8.00)^2 \text{ g/mol}$$

$$\underline{M_1 = 64 \text{ g/mol}}$$

11. The difficulty here is that we don't have a formula for the combustion; however, we do know that each mole of carbon in the original sample must produce one mole of carbon dioxide. It seems logical to begin by calculating the moles of carbon dioxide formed.

Since the gas is measured at STP, use the molar volume to determine the moles of CO_2.

$$\text{moles } CO_2 = \frac{67.2 \text{ L}}{22.41 \text{ L/mol}}$$

$$\text{moles } CO_2 = 2.999 \text{ moles}$$

Therefore, the combustion of 0.500 moles of hydrocarbon produces 2.999 moles of carbon dioxide. How many moles of carbon dioxide would be produced by combusting 1.00 mole of hydrocarbon?

mol CO_2 /mol of cpd = 2.999 mol CO_2 / 0.500 mol cpd

mol CO_2 /mol of cpd = 6.00 mol CO_2 /mol cpd

But each mole of CO_2 produced indicates one mole of carbon present in the compound. It must also be true that

mole C /mole of cpd = 6.00 mol C /mol cpd

The empirical formula tells us that there are three moles of hydrogen for each mole of carbon, so the molecular formula must be

$$\underline{C_6H_{18}}$$

12. Since the container is initially at standard pressure, the pressure on the oxygen alone must be 1.00 atm. When the gases are mixed, according to Dalton's law this becomes the partial pressure of the oxygen. Similarly, since the volume of the hydrogen is identical with that of the oxygen under similar conditions, the initial pressure of the hydrogen must also be 1.00 atm. If the temperature is left unchanged, the new total pressure must simply be the sum of the partial pressures, that is,

$$P_{total} = P_{O_2} + P_{H_2} = 1.00 \text{ atm} + 1.00 \text{ atm}$$

And so the total pressure must be 2.00 atmospheres.

13. a. A data table is useful here, but first use Dalton's law to find the total moles of gas.

$$\text{total moles of gas} = 0.010 \text{ mol} + 0.030 \text{ mol}$$
$$= 0.040 \text{ mol}$$
$$\text{temperature} = 383 \text{ K}$$
$$\text{pressure} = 1.00 \text{ atm}$$
$$\text{volume} = ?$$

Substitute these values into the ideal gas law and solve for V

$$V = \frac{nRT}{P} = \frac{(0.040 \text{ mol})(0.08206 \text{ L atm/mol K})(383 \text{ K})}{1.00 \text{ atm}}$$

V = 1.3 liters

b. Begin by balancing the equation

$$2 \text{ CH}_4(g) + 3 \text{ O}_2(g) \rightarrow 2 \text{ CO}(g) + 4 \text{ H}_2\text{O}(g)$$

By inspection of the equation, it is apparent that 0.030 mole of oxygen will require 0.020 moles of methane for complete reaction. Since the amount of

227

methane is insufficient, methane must be the limiting reagent.

$$\text{mol CO formed} = \text{mol CH}_4 \times \frac{2 \text{ mol CO}}{2 \text{ mol CH}_4}$$

mol CO formed = 0.010 mole

c. Let's review the moles of gas present at the end of the reaction:

0.010 moles of CO formed

0.0150 moles of oxygen didn't react

From the balanced equation, the moles of water formed will be twice the moles of carbon monoxide, so there will be 0.020 moles of water formed.

Total moles of gas

$$= 0.015 \text{ mol O}_2 + 0.010 \text{ mol CO} + 0.020 \text{ mol H}_2\text{O}$$

Total moles of gas = 0.045 moles

Now write the data table for the system when the reaction is completed.

```
total moles of gas = 0.045 moles
temperature       = 383 K
pressure          = ?
volume            = 1.257 L
```

Substitute these values into the ideal gas law and solve for P

$$P = \frac{nRT}{V} = \frac{(0.045 \text{ mol})(0.0821 \text{ L atm/mol K})(383 \text{ K})}{1.26 \text{ L}}$$

P = 1.1 atm

PRACTICE TEST (50 minutes)

1. A sample of oxygen gas has a volume of 2.50 dm^3 at standard pressure. Calculate the volume (in dm^3) this same sample would occupy at 50.0 kPa.

2. Calculate the number of molecules of nitrogen in a pure sample of this gas having a volume of 0.210 liters at STP.

3. If the temperature and pressure are kept constant during the process, how many liters of titanium(IV) chloride will be produced when 10.0 liters of chlorine reacts with excess titanium according to the equation

$$Ti(s) \quad + \quad 2\ Cl_2(g) \quad \rightarrow \quad TiCl_4(g)$$

4. A sample of nitrogen gas occupies a volume of 27.9 mL at a temperature of -80.0°C. The gas is heated until it occupies a volume of 85.5 mL. Assuming the pressure is constant, determine the new temperature of this sample in Celsius degrees.

5. A sample of oxygen gas having a volume of 56.0 milliliters is collected by water displacement at a temperature of 20°C and a pressure of 710.0 mmHg. Determine the volume of this oxygen sample as a dry gas at STP. (The vapor pressure of water is 17.5 mmHg at 20°C.)

6. If a certain gas sample has a volume of 30.0 cubic feet at a pressure of 500.0 mmHg, what is the volume of this same sample in cubic feet at 2.0 atmospheres?

7. Calculate the molar mass of an unknown gas if at STP a sample of this gas has a density of 2.86 grams/liter.

8. Calculate the pressure in atmospheres which will result in a sealed container having a volume of 23.5 milliliters if 1.00 grams of BCl_3 gas is the only substance in the container and the temperature is 273°C.

9. Under the same conditions of temperature and pressure, methane gas (CH_4) diffuses through a porous barrier at a rate 2.3 times as fast as does a certain unknown gas. What is the molar mass of the unknown gas?

10. A certain gaseous compound, C_xH_y, contains only the elements carbon and hydrogen. When a sample of this compound is combusted, the only products are 5.38 L of carbon dioxide and 8.06 L of water, both measured at STP. (a) What is the empirical formula of the unknown compound? (b) If the volume of the original unknown gas was measured to be 2.69 L at STP, what is the molecular formula of the unknown compound?

11. A sample of unknown liquid is placed into a weighed, evacuated flask of known volume at a temperature high enough to vaporize all of the liquid. The temperature is held constant and the pressure in the flask is measured. The flask is weighed again to determine the mass of the unknown liquid. Using the data below obtained by this procedure, calculate the molar mass of this unknown liquid.

mass of the empty flask	35.364 grams
volume of the flask	35.0 mL
pressure in the flask	381 mmHg
mass of the flask and the unknown	35.451 grams
temperature	$100.0°C$

12. Hydrogen gas and oxygen gas are reacted to form water according to the balanced equation

$$2 \; H_2(g) \; + \; O_2(g) \; \rightarrow \; 2 \; H_2O(g)$$

If 20.0 liters of hydrogen is reacted with 30.0 liters of oxygen, calculate the total volume of gas at the end of the reaction. You may assume that the temperature is high enough that all of the water produced remains in the gas state, and that all gas volumes are measured at the same temperature and pressure.

COMMENTS ON ARMCHAIR EXERCISES

1. Did you recognize that this is simply a practical application of Charles's law?

2. When you drop the tablet in the water, a gas is generated. Do you know that the gas is? As the

gas is generated in the confined space in the can, what happens to the pressure? Does this suggest an explanation for the small explosion?

3. Do you see a relationship between the height of the column of water and the pressure on the water as it escapes? Does this suggest a similar relationship should occur in the atmosphere?

4. The key here is to ask how many molecules collide with the satellite. Even though the effective energy of each molecule may be high, if there are few collisions, there is little energy exchange.

CONCEPT TEST ANSWERS

1. 760.00 mmHg, 101.325 kPa   2. pressure
3. be half as great   4. pressure
5. absolute or kelvin temperature
   (Not just temperature!)
6. double   7. $-273.15°C$
8. molecules   9. double
10. 22.414 L, standard molar volume
11. 0.082057 L·atm/mol·K
12. standard temperature and pressure
13. it will double   14. 610 mmHg
15. partial pressures
16. vapor pressure of water   17. very large
18. temperature   19. increased
20. c. low temperature, high pressure
21. a. O_2 b. CCl_4

PRACTICE TEST ANSWERS

1. (L2) 5.07 dm^3 2. (L3) 5.64×10^{21}
3. (L6) 5.00 liters   4. (L2) $319°C$
5. (L5 & L7) 47.5 mL   6. (L2) 9.9 ft^3
7. (L4) 64.1 g/mol   8. (L4) 16.3 atm
9. (L9) 85 g/mol   10. (L3) CH_3, C_2H_6
11. (L4) 150 g/mol
12. (L6) 40.0 liters of gas

CHAPTER 13
BONDING AND MOLECULAR STRUCTURE:
INTERMOLECULAR FORCES, LIQUIDS, AND SOLIDS

CHAPTER OVERVIEW

This chapter continues the review of the three states of matter. The properties of liquids and solids are quite different from those of gases, because of the stronger intermolecular forces. Depending on the situation, the attractions in liquids and solids may involve either molecules or ions, and the molecules may have a permanent dipole or one that is induced by the presence of other charged species. In order to understand the intermolecular forces in solids and liquids, it is necessary to consider the various types of interactions that are possible.

Hydrogen bonding is an especially important example of intermolecular interaction. This attraction is largely responsible for the unusual properties of water. Since this solvent is so fundamental to life as we know it, hydrogen bonding occupies a special place in the discussion. Other molecules also undergo hydrogen bonding, although none of them is as important as water.

The kinetic molecular theory, which was discussed earlier in connection with the behavior of gases, also helps to explain many of the properties of liquids and even solids. Both vaporization and melting are best understood in terms of this theory.

In order to describe the regular structure of crystal lattices, it is helpful to consider a simple repeating unit, called the unit cell. Once the unit cell idea is understood, a number of other properties can be related to it, including atomic radius, density, and even Avogadro's number. These relationships can also be extended to ionic solids as well as molecular solids.

Although a phase diagram may look simple, it summarizes a great deal of data about the conversions among the three states of matter.

LEARNING GOALS

1. Intermolecular attractions include ion-dipole, dipole-dipole, dipole-induced dipole, and induced dipole-induced dipole interactions. You should understand the way in which these forces arise and be aware of their relative strengths. Hydrogen bonding is an especially important example of dipole-dipole interactions, and you should understand why the existence of hydrogen bonds contributes to the unusual properties of water. (Sec. 13.2)

2. You should understand liquid behavior, including such properties as vaporization, vapor pressure, boiling, the critical point, surface tension, and viscosity. (Sec. 13.3)

3. Be able to recognize the different types of metallic and ionic solids, as well as the common cubic unit cells. Your knowledge of crystalline lattices should be sufficient to allow you to use information about the density and type of unit cell to determine the cell dimensions or atom radius, as well as the reverse process, determining the cell type from knowledge of the density and ionic radii. (Sec. 13.4)

4. Several common substances, like water, carbon dioxide, graphite and diamond, form either molecular or network solids. You should be able to recognize when this type of solid is possible. (Sec. 13.5)

5. Be familiar with some of the physical properties of solids, including melting point, heat of fusion, sublimation, and heat of vaporization. (Sec. 13.6)

6. A phase diagram shows how the various possible phases of a substance are affected by temperature and pressure. If you are given the phase diagram for a substance you should be able to use it to discuss the effects of temperature and pressure changes on the behavior of that substance, including what happens at the critical point. You should also be able to draw a phase diagram when given appropriate information regarding a pure substance. (Sec. 13.7)

ARMCHAIR EXERCISES

1. Water is commonly used to put out fires. Look carefully at the properties of water (see the table on page 604) to determine if water has any special characteristic that makes it an excellent liquid for extinguishing fires.

2. After water has been boiling for several minutes, what is the gas that fills the bubbles that you see forming in the boiling water?

3. Oil for automobile engines is rated on the basis of its viscosity. For example, 10 weight oil is lower in viscosity than 40 weight oil. At one time, before the modern lubricants made it unnecessary, people who lived in very cold climates would change their oil to a lighter weight in the winter. Why was this necessary?

4. In her book, *Chemistry for Every Kid*, Janice VanCleave describes a simple experiment that demonstrates some unexpected behavior by water. Four holes are punched in a straight line near the bottom of a medium or large paper or styrofoam cup. Then the cup is placed over a sink and filled with water. As the water flows out of the four holes, it can be pinched together with the fingers to form a single stream, that remains united even after the fingers are removed. Why do the four streams of water unite into one?

5. It is possible to make water boil without apparently using heat. If you fill a plastic syringe about one-third full of moderately warm (about 65°C) water, then close the end of the syringe with a cork, and pull out the plunger slowly, the water will begin to boil. (BE CAREFUL NOT TO USE WATER THAT IS TOO HOT OR YOU MAY BURN YOURSELF.) The water in the plunger will begin to bubble as though it were boiling.
   Is the water really boiling? Why is this phenomena observed?

6. Pluto, the outermost planet in our solar system, is thought to be so cold that it may have frozen lakes of methane. (Methane is a simple hydrocarbon that you learned about in Chapter 11.) Would you

guess that you can skate on a frozen lake of methane the way you skate on frozen water? (Of course, this is just a theoretical question, since the lack of air, the low temperatures, and the inaccessibility of Pluto prevent us from actually testing your answer.)

CONCEPT TEST

1. At normal temperatures and pressures, most of the elements are in the _____ state.

2. _____ is the state of matter demonstrates the largest volume change due to a change in pressure.

3. Which type of intermolecular force is strongest?
a. ion-dipole b. induced dipole-induced dipole
c. dipole-induced dipole d. dipole-dipole

4. When water molecules are bound to a metal ion, the ion is said to be _____.

5. As the metal ion radius becomes larger, the enthalpy of hydration becomes _____ exothermic. (choose from less or more)

6. The saying that "like dissolves like" means that polar molecules are more likely to dissolve in a(n) _____ solvent.

7. When comparing the boiling points of HF, HCl, HBr, and HI, the boiling point of HF is much higher than might be predicted because of _____.

8. What three elements are likely to cause hydrogen bonding when attached to hydrogen? _____

9. The degree to which the electron cloud of an atom or nonpolar molecule can be distorted by an external electric charge depends on the atom's or molecule's _____.

10. Ice is _____ than liquid water. (Choose from less dense, more dense, or the same density.)

11. Vaporization is always a(n) _____ process. (choose from endothermic or exothermic.)

12. The curve of vapor pressure versus temperature on a phase diagram comes to a abrupt halt at the combination of critical temperature and critical pressure called the _____.

13. When liquid/vapor equilibrium has been established, the pressure exerted by the vapor is called the _____.

14. The energy required to break through the surface of a certain liquid sample or to disrupt a drop of that liquid so that the material spreads out as a film is called _____.

15. When a liquid is placed in a tube, the interaction among the adhesion of the liquid and the walls of the tube, the cohesion between the liquid molecules, and the force of gravity produces a characteristic concave or convex curve of the liquid surface called the _____.

16. The smallest, repeating unit that has all of the symmetry characteristics of the way that the atoms are arranged in a crystal lattice is called the _____.

17. Each atom on the corner of a cubic unit cell will be shared by total of _____ unit cells; each atom in a face of a cubic unit cell will be shared by _____ unit cells, and each atom in the center of a cubic unit cell will be shared by

_____ unit cells.

18. The _____ of a solid is defined as the temperature at which the crystal lattice collapses and the solid is converted to a liquid.

19. When a solid is directly transformed into a gas without becoming a liquid, the process is called

_____.

20. Each line in a phase diagram represents the conditions of temperature and pressure at which equilibrium exists between _____.

21. The conditions of temperature and pressure at which three different phases can be in equilibrium is called the _____.

22. Substances can sublime only at temperatures and pressures that are below the _____.

PRACTICE PROBLEMS

1. (L1) in each of the following cases, determine what type of intermolecular force is involved and arrange the examples in order of increasing strength of attraction. a. NaF . . . H_2O, b. Rn . . . Rn, and c. NO . . . NO.

2. (L2) When a pure liquid is boiling, the temperature doesn't change, even though heat is constantly being applied to the system. Can you explain this in terms of the kinetic molecular theory?

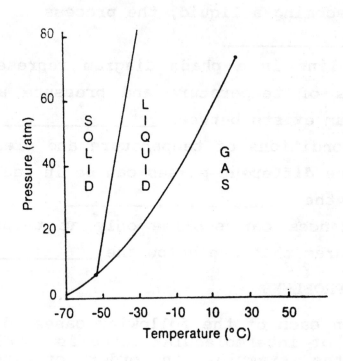

Figure 13.1 Rough Phase Diagram for CO_2

3. **(L6)** The above phase diagram is based on the following physical property data for carbon dioxide:

sublimation point $-78.5°C$ at 1.00 atm
triple point $-56.5°C$ at 5.11 atm
critical point $31.1°C$ at 72.9 atm

Use this information to answer the questions on the next page   regarding the behavior of carbon dioxide.

238

a. Which is more dense, solid CO_2,
   liquid CO_2, or do both states
   have the same density?   _____

b. What is the normal boiling point
   for carbon dioxide?   _____

c. Is CO_2 a gas, liquid, or solid
   at $-20°C$ and 65 atm?   _____

d. Approximately what pressure
   would be necessary to liquify
   CO_2 at $35°C$?   _____

4. (L5) How much heat will be required to
completely vaporize 1.00 liter of butane at its
boiling point. For liquid butane, the density is
0.601 g/mL, the molar mass is 58.13 g/mol, and the
heat of vaporization is 24.3 kJ/mol.

5. (L2) Match the liquid property in the left hand
column with the appropriate description from the
right hand column by writing the appropriate letter
in the space next to the property on the left.

_____ surface tension   a. causes water to
                                     rise in a tube

_____ critical point   b. energy required
                                     to overcome the
                                     attractive forces on
                                     the surface molecules

_____ capillary action   c. resistance to flow

_____ viscosity   d. highest
                                     temperature and
                                     pressure at which a
                                     gas can be liquified
                                     by application of
                                     pressure

6. **(L5)** Carbon disulfide is an unpleasant smelling liquid that has a vapor pressure of 298 mm Hg at $20°C$. If the atmosphere of a room having a volume of 30.0 liters is saturated with this vapor at $20°$, how many grams of CS_2 are contained in the air?

7. **(L4)** Complete the table below by matching the type of solid from the left-hand column with the substance from the right-hand column that is an example of that type of solid.

_____ metallic solid   a. diamond
_____ ionic solid   b. I_2
_____ molecular solid   c. pure silver
_____ network solid   d. CsF

8. **(L3)** The density of vanadium is 6.11 g/cm^3, and it crystallizes with a body-centered cubic unit cell. (a) Calculate the length (in nanometers) of a side of the unit cell for vanadium. (b) Based on this information, what is the radius of a vanadium atom?

PRACTICE PROBLEM SOLUTIONS

1. Sodium fluoride is ionic and water is polar, so the intermolecular forces involved will be ion-dipole. These are very strong, and since no other ionic attractions are present, this will be the

strongest attraction.

Radon gas is nonpolar, so the only attractive forces will be induced dipole-induced dipole. These forces provide the weakest type of attraction.

Nitrogen oxide is a dipolar molecule, so the attraction is dipole-dipole, which is normally intermediate in strength between the previous two cases. Thus the order is

   b. Rn. . Rn < c. NO. . NO < a. NaF. . H_2O

2. Even though energy is constantly being added to the boiling liquid, it is also constantly being removed by the molecules that are escaping from the surface of the liquid.

3. a. solid CO_2 is more dense than liquid CO_2
   b. Liquid and solid carbon dioxide cannot be in equilibrium at 760 mm Hg, and so this compound doesn't have a normal boiling point.
   c. liquid
   d. This temperature is above the critical temperature, and so carbon dioxide cannot be changed into a liquid by the application of pressure.

4. First, use the density equation to determine how many grams of butane are contained in 1.00 Liter.

$$mass = density \times volume$$

$$= 0.601 \text{ g/mL} \times 1.00 \text{ L} \times 1000 \text{ mL/L}$$

$$= 601 \text{ grams}$$

Next, divide by the molar mass to determine the number of moles.

$$moles = \frac{\text{grams of butane}}{\text{molar mass}}$$

$$= \frac{601 \text{ grams}}{58.13 \text{ g/mole}}$$

= 10.34 moles of butane

Finally, determine the energy required.

q = 10.34 mol x 24.3 kJ/mol

q = 251 kJ

5. surface tension   b. energy required to overcome the attractive forces on the surface molecules

critical point   d. highest temperature and pressure at which a gas can be liquified by application of pressure

capillary action   a. causes water to rise in a tube

viscosity   c. resistance to flow

6. Use the ideal gas law to calculate the amount of carbon disulfide that would be necessary to exert the observed vapor pressure.

$$\text{moles of } CS_2 = \frac{PV}{RT} = \frac{(298 \text{ mm}/760 \text{ mm/atm})(30.0 \text{ L})}{(0.08206 \text{ L·atm/mol·K})(293.2 \text{ K})}$$

moles of CS_2 = 0.489 moles

mass of CS_2 = 0.489 moles x 76.14 g/mol

mass of CS_2 = 37.2 grams

7. metallic solid   c. pure silver
ionic solid   d. CsF
molecular solid   b. I_2
network solid   a. diamond

8. a. First determine how many atoms of vanadium are present in a single unit cell. There are eight corner atoms, each shared by seven other unit cells, and one atom in the center that isn't shared with any other unit cell. Thus, the total is

8 corner atoms x 1/8 atom/site = 1 atom
1 central atom (not shared) = 1
Total atoms per unit cell = 2

This allows you to calculate the mass of a single unit cell:

$$\text{unit cell mass} = \frac{2 \text{ atom/cell} \times 50.94 \text{ g/mole}}{6.022 \times 10^{23} \text{ atom/mole}}$$

unit cell mass = 1.6924×10^{-22} grams

Use the density to determine the volume of a unit cell.

$$\text{unit cell volume} = \frac{1.69 \times 10^{-22} \text{ grams}}{6.11 \text{ g/cm}^3}$$

unit cell volume = 2.77×10^{-23} cm^3

Take the cube root of the volume to find the length of a side.

$$\text{side of unit cell} = \sqrt[3]{2.77 \times 10^{-23} \text{ cm}^3}$$

$$= 3.03 \times 10^{-8} \text{ cm}$$

Convert to nanometers

$$\text{Length of side (nm)} = \frac{3.03 \times 10^{-8} \text{ cm} \times 1 \times 10^{9} \text{ nm/m}}{100 \text{ cm/m}}$$

Length of side = 0.303 nm

b. For a body-centered cubic cell, the contact between atoms occurs along the diagonal of the cube. The length of this diagonal is the square root of 3 times the length of the side of the cell.

Since that distance is equivalent to four times the radius of the atom

$$\text{atomic radius} = \frac{s \sqrt{3}}{4} = 0.303 \text{ nm} \times 0.4330$$

atomic radius = 0.131 nm

PRACTICE TEST (35 Minutes)

1. Arrange the following types of intermolecular attractions in order of increasing magnitude of the force exerted:

ion-dipole   induced dipole-induced dipole
hydrogen bonding   ion-ion

2. Arrange the following ions in order of increasing energy of hydration: Ca^{2+}, Cs^+, and Ba^{2+}

3. In the northern parts of the United States, a snowfall will begin to disappear from the ground in a few days, even though the temperature never reaches the melting point of water. Can you explain this?

4. Complete the table below by matching the type of solid from the left-hand column with the substance from the right-hand column that is an example of that type of solid.

_____ metallic solid   a. silicates
_____ ionic solid   b. an iron bar
_____ molecular solid   c. KI
_____ network solid   d. ice

5. The standard enthalpy of formation of liquid carbon tetrachloride is -135.44 kJ/mole, and the standard enthalpy of formation of gaseous carbon tetrachloride is -102.9 kJ/mole, what is the standard molar enthalpy of vaporization of CCl_4.

6. Carbon tetrachloride, CCl_4, is a toxic organic liquid that vaporizes rather readily. If an open container of carbon tetrachloride is placed in a sealed room at $20°C$ and allowed to come to equilibrium, it is observed that 18.4 grams of carbon tetrachloride are present in the 24,000 milliliters of air in the room. Based on this information, what is the vapor pressure of carbon tetrachloride in units of mm Hg?

7. Metallic silver has a density of 10.5 g/cm^3 and crystallizes with a face-centered cubic unit cell.

(a) Calculate the length of a side of the unit cell (in nanometers) for silver. (b) Based on this information, what is the radius of a silver atom?

8. PHYSICAL PROPERTY DATA FOR CARBON MONOXIDE

normal melting point 68.09 K
normal boiling point 81.65 K
triple point   68.10 K at 115.4 mm Hg

Using the information provided above, draw a rough phase diagram for carbon monoxide, and use your diagram to answer the questions below. You should label all axes, and indicate the stable state (gas, liquid, or solid) in each region of the diagram.

a. Is carbon monoxide a gas, a solid,
or a liquid at 71°K and 600 mm Hg?   _____

b. Which is more dense, solid carbon
monoxide, liquid carbon monoxide,
or do both states have the same density?_____

c. Will carbon monoxide sublime at a
pressure of 1 atm? If not, indicate
a pressure at which it will sublime.   _____

d. If the critical point for CO is at
132.9K and 34.5 atmospheres, approximately
what pressure would be necessary to
liquify CO at a pressure of 35 atm?   _____

COMMENTS ON ARMCHAIR EXERCISES

1. Water has the highest heat of vaporization. That means that when it is sprayed on a fire, a large amount of energy will be used to vaporize the water. This should cool down the combustible materials enough to prevent re-ignition.

2. The bubbles are filled with water vapor.

3. As the temperature became colder, the viscosity of the oil decreased. At very low temperatures, the oil began to have such a high viscosity that the engine parts could no longer move. This would

make it very hard to start the car on a cold morning. You can now buy multiviscosity oils, and so this type of problem is much less common.

4. Something is obviously attracting the four streams of water together. What force might be responsible for this? What do you know about intermolecular attraction in water?

5. When you pull out the plunger, what happens to the pressure inside the syringe? As the pressure changes, what happens to the boiling point of the water?

6. As noted in the text, water is very unusual in that the liquid at the freezing point is more dense than the solid. This is what makes it possible to melt the ice by applying pressure with the skate blade. You should guess that for methane the solid is more dense than the liquid (it is!). Will solid methane melt when pressure is applied by a skate blade?

CONCEPT TEST ANSWERS

1. solid
2. gas
3. a. ion-dipole
4. hydrated
5. less
6. polar
7. hydrogen bonding
8. N, O, and F
9. polarizability
10. less dense
11. endothermic
12. critical point
13 equilibrium vapor pressure
14. surface tension
15. meniscus
16. unit cell
17. eight, two, and one. Notice that the question asks for the total number of unit cells sharing the atom.
18. melting point
19. sublimation
20. the two phases on either side of the line
21. triple point
22. triple point

PRACTICE TEST ANSWERS

1. (L1) induced dipole-induced dipole < hydrogen bonding < ion-dipole < ion-ion
2. (L1) $Cs^+ < Ba^{2+} < Ca^{2+}$
3. (L5) The solid snow is directly converted into

water vapor by the process of sublimation.
4. (L3) metallic solid   b. an iron bar
    ionic solid   c. KI
    molecular solid   d. ice
    network solid   a. silicate
5. (L2) +32.5 kJ/mole
6. (L2) 91 mm Hg
7. (L3) a. 0.409 nanometers b. 0.145 nm

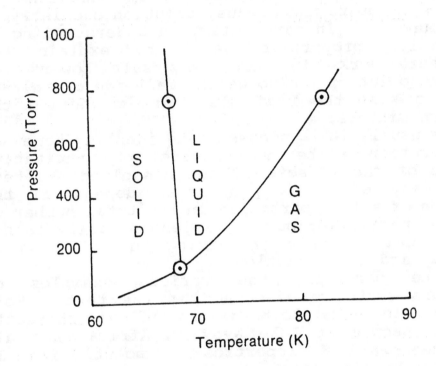

Figure 13.2 Approximate Phase Diagram for
           Carbon Monoxide

8. (L6) The phase diagram is shown above.
   a. liquid
   b. liquid
   c. no, below 115.4 mm Hg
   d. Can't be done; this is above the critical
      temperature.

CHAPTER 14
SOLUTIONS
AND THEIR BEHAVIOR

CHAPTER OVERVIEW

The practical chemistry of our common lives is the chemistry of solutions. From reactions in living systems to the oceans, solution chemistry is all around us. In particular, an understanding of colligative properties helps to explain how antifreeze works in cars, why salt lowers the freezing point of ice so we can make ice cream, and helps to keep the road clear of ice during the northern winters.

As usual, it is necessary to identify some new concentration units to provide a quantitative measure of the effects of the dissolved solutes. Since many of these properties depend on the fraction of solute particles in the total number of solution particles, our old friend molarity is not usually helpful here, requiring the addition of molality and mole fraction.

Note carefully the various examples of solutions that are discussed in the text. Soft drinks, cold packs, scuba divers and even the death of 1700 people at Lake Nyos in Africa, are all explained by the properties of solutions. In addition, boiling point elevation or freezing point depression is used to experimentally determine the molar mass of solutes. An understanding of solutions is also necessary to explain osmosis, which is very important in biology.

The final section, which describes colloids, also has many practical applications, ranging from detergents to foods.

LEARNING GOALS

1. If necessary, you should review the definitions of solute, solvent, solution, and molarity. This will prepare you to understand the new concentration units, including mole fraction,

molality, and weight percent. (Sec. 14.1)

2. Understand the factors that affect solubility for various types of solutions and be familiar with the use of Henry's law and le Chatelier's principle to predict the effects of temperature and pressure on solubility. (Sec. 14.2)

3. You should understand how a nonvolatile solute can affect solvent vapor pressure and also be able to use Raoult's law to calculate the magnitude of this effect. Raoult's law can also be used to determine the molar mass of nonvolatile solvents by measuring the decreased vapor pressure of a solvent. (Sec. 14.3)

4. Freezing point depression and boiling point elevation due to the presence of dissolved solutes are frequently encountered applications of the theories discussed in this chapter. You should be able to work problems based on freezing point and boiling point change for both nonionizing solutes as well as solutes that ionize to produce more than one mole of particles per mole of solute. (Sec. 14.3)

5. Recognize that osmotic pressure is similar to the colligative properties studied previously and be able to do calculations based on the equation that relates osmotic pressure and molarity. (Sec. 14.3)

$$\Pi = MRT$$

6. Colloidal dispersions are commonly encountered in everyday life. Recognize how colloids differ from true solutions, and be able to identify some common emulsifying agents, including soaps and detergents. (Sec. 14.4)

ARMCHAIR EXERCISES

1. Two beakers, one half filled with pure water and the other half filled with a concentrated salt solution, are placed in a closed container and allowed to stand for a long time. When the beakers are removed and examined, it is found that the beaker that contained the pure water is now empty and the beaker that contained the solution has even

more liquid than before. Can you explain this observation?

2. Everyone knows that oil and water don't mix. If you add some cooking oil to water and shake the mixture, in a short time the two layers will separate again. (Does the rule of thumb that like dissolves like help you to understand this?) Suppose that you add a little detergent to the mixture and shake again. Now the result is a mixture that looks homogeneous. What did the detergent do to change the situation?

3. Muriel Mandell suggests a solution experiment in her book, *Simple Kitchen Experiments*. She directs you to cut three identical sized one inch cubes of raw potato, then place one cube in a tap water, one in a dilute solution of salt in water, and one in a solution prepared by adding a large handful of salt to the water. When the cubes are examined after an hour, it is found that the cube that had been in the concentrated salt solution is now much smaller, and the cube that was in the tap water will be a little bigger. Can you explain this result?

CONCEPT TEST

1. A homogeneous mixture of two or more substances in a single phase is called a(n) _____.

2. Properties that ideally depend only on the number of solute particles per solvent molecule and not on the nature of the solute or solvent are called _____ properties.

3. _____ is the concentration unit defined as the number of moles of solute per kilogram of solvent.

4. The sum of all of the mole fractions of all components in a solution must always equal a value of _____.

5. _____ is the concentration unit

defined as the mass of solute per 100 grams of solvent.

6. _____ is the term used to describe the maximum amount of material that will dissolve in a given amount of solvent at a given temperature to produce a stable solution.

7. When a solution temporarily contains more solute than the saturation amount, the solution is said to be _____.

8. If two liquids mix to an appreciable degree to form a solution they are said to be _____.

9. The general rule of thumb for predicting solubilities is that like dissolves like. For instance, polar solvents are most likely to dissolve _____ solutes, and nonpolar solutes are most likely to dissolve _____ solutes.

10. According to Henry's law, the solubility of a gas in a liquid is directly proportional to the _____ of the gas.

11. The solubility of gases in liquid solvents always _____ (choose from increases or decreases) when the temperature increases.

12. The solubility of most solids will increase when the temperature increases because this is usually an _____ (choose from exothermic or endothermic) process.

13. The lowering of the vapor pressure of a solution compared to the vapor pressure of the pure solvent is proportional to the _____ of the solute.

14. The boiling point of a solution is _____ than that of the pure solvent, whereas the freezing point of a solution is _____ than that of the pure solvent. (Choose your answers from higher or lower.)

15. As a solution freezes, solvent molecules are removed from the liquid phase and the solution becomes more concentrated, causing the freezing point of the remaining solution to _____.

16. In normal osmosis, molecules spontaneously move across a semipermeable membrane to a solution having a _____ (higher or lower?) concentration of solvent molecules.

17. Colloidal suspensions represent a state intermediate between a _____ and a suspension.

18. Colloidal dispersions of one liquid in another, such as oil in water, are called _____.

19. Soaps have a nonpolar, hydrophobic end that is soluble in nonpolar materials, like grease, and a polar, hydrophilic end that is soluble in _____.

20. Water that contains a high concentration of ions, like Ca^{2+} and Mg^{2+}, that for a precipitate with soap is said to be _____ water.

STUDY HINTS

1. At first glance, molarity and molality seem to be very similar, but notice that molarity involves the volume (liters) of solution and molality involves the mass (kilograms) of solvent. Learn these differences carefully, so that they won't be overlooked in the pressure of an examination.

2. Remember that concentration units are extensive

properties, that is, the concentration value is independent of the amount of solution. In some problems it will be necessary to assume a mass or volume of solution. This is permitted as long as no values are given that limit the quantity of solution and as long as only one such assumption regarding the quantity of solution is made in a given problem.

3. Even after you are convinced that it is mathematically correct to assume some amount of solvent, solute, or solution in some of these problems, you may still feel a little confused about which quantity is the best to choose. Unfortunately there is no hard and fast rule you can memorize, but in most cases it is simplest to assume some amount of the substance in the denominator of the definition of the concentration unit that has a known value. If you know molality, assume a mass of solvent; if you know molarity, assume a volume of solution, and if you know the weight percent, assume a mass of solution. This plan should provide you with some guidance until you become more familiar with this type of problem.

PRACTICE PROBLEMS

1. **(L1)** The density of a 4.10 molal solution of NaCl is 1.120 grams/mL. Use this information to calculate (a) the weight percent and (b) the molarity of this solution.

2. (L2) The rule that "like dissolves like" often provides practical help when preparing solutions. Suppose only two solvents are available, toluene (a relatively nonpolar organic liquid) and water. For each of these substances, predict which solvent will be most likely to dissolve the material.

a. LiCl
c. benzene

b. acetic acid
d. methane, CH_4

3. (L4) a. Suppose that you have 1.00 mole of each of the following compounds: $BaCl_2$, CsCl, and $C_2H_4(OH)_2$ (ethylene glycol). Which will cause the greatest decrease in the freezing point of a 10.0 liter sample of water? Why is this true?

b. Suppose that you have 1.0×10^2 grams of each of the above compounds. Now which sample will cause the greatest decrease in the freezing point of the water? Explain this.

4. (L3) Consider a solution consisting of 50.0 grams of benzene (78.12 g/mole) and 50.0 grams of n-octane (molar mass = 114.2 g/mole). At 20°C the vapor pressure of pure benzene is 95.2 mm Hg, and the vapor pressure of pure n-octane is 14.1 mmHg. Use Raoult's Law to calculate the equilibrium vapor pressures of benzene and n-octane over this solution.

5. **(L5)** If 95.2 grams of a protein sample is dissolved in enough water to produce 1950 milliliters of solution, the resulting osmotic pressure is 0.0675 atmospheres at 25 degrees Celsius. Calculate the molar mass of this protein.

6. **(L4)** Calculate the freezing point of the solution in an automobile radiator that is 30.0% by weight ethylene glycol, $C_2H_4(OH)_2$. (Remember that the freezing point depression constant for water is $-1.86°C/m$ for water.)

7. **(L6)** A number of different terms are used to describe colloidal dispersions. For each term given, identify what is special about that type of colloidal dispersion, or if the term does not refer to a colloid, explain how it differs from a colloid.

a. emulsion

b. suspension

c. sol

d. gel

8. (L2) In some areas the well water has an odor like rotten eggs due to dissolved hydrogen sulfide gas. If the value of the Henry's law constant for H_2S is 2.6×10^{-4} molal/mm Hg at $25°C$, and the concentration of hydrogen sulfide in the water was 0.050 molal, what was the pressure in atmospheres of the hydrogen sulfide gas when it was dissolved underground?

PRACTICE PROBLEM SOLUTIONS

1. Since we are given the molal concentration, it will probably be best if we assume a given mass of solvent. This will be most directly useful with the definition of molality. We will begin by assuming we have 1.00 kilograms of water.

a. First find the total moles of NaCl
moles of solute = kg of solvent x molality
= 1.00 kg x 4.10 m
= 4.10 moles

Next find the mass of NaCl
mass of NaCl = 4.10 moles x 58.44 g/mole
= 239.6 grams

Total mass of solution
= mass of NaCl + mass of water
= 239.6 g + 1000. g
= 1239.6 g

$$\text{Percent NaCl} = \frac{239.6 \text{ g} \times 100}{1239.6 \text{ g}}$$

Percent NaCl = 19.3 %

b. In order to use the molarity equation, you must know the volume of the solution. This can be determined from the density and the total mass of the solution.

Density = mass/volume

$$\text{Volume of solution} = \frac{\text{mass}}{\text{density}} = \frac{1239.6 \text{ g}}{1.120 \text{ g/mL}}$$

Volume of solution = 1106.8 mL

Now you know the volume of solution and the moles of solute, and so you can use the molarity equation.

$$\text{molarity} = \frac{\text{moles of solute}}{\text{liters of solution}}$$

$$\text{molarity} = \frac{4.10 \text{ moles}}{1.1068 \text{ liters}}$$

<u>molarity = 3.70 M</u>

2. Lithium chloride and acetic acid are both polar solutes, and so they are more likely to be soluble in a polar solvent like water. Benzene and methane are relatively nonpolar, and so are expected to be more soluble in toluene, a nonpolar solvent.

3. a. Barium chloride and cesium chloride are both ionic compounds; assuming that the dissociation is essentially complete, one mole of barium chloride will produce three moles of ions, and one mole of cesium chloride will produce two moles of ions. The ethylene glycol doesn't ionize, and so it will produce only one mole of solute particles. Since freezing point depression depends on moles of solute particles, the barium chloride will cause the greatest change in the freezing point.

b. The molar masses of the three compounds are not the same, so 100 grams does not represent the same number of moles of each. The explanation is probably clearest if based on a data table, such as that below.

compound	molar mass	moles of cpd per 100 g	moles of ions per 100 g
$C_2H_4(OH)_2$	62.1 g/mole	1.61 moles	1.6 moles
CsCl	168	0.595	1.2
$BaCl_2$	208	0.481	1.4

Under these conditions, ethylene glycol produces the greatest number of ions in solution, and so it will cause the greatest freezing point depression.

4. First, calculate the moles of each compound:

$$\text{moles of benzene} = \frac{50.0 \text{ g}}{78.12 \text{ g/mole}} = 0.6400 \text{ moles}$$

$$\text{moles of octane} = \frac{50.0 \text{ g}}{114.2 \text{ g/mole}} = 0.4378 \text{ moles}$$

Calculate the mole fraction of each component:

mole fraction of benzene = X_b

$$= \frac{0.6400 \text{ mol}}{0.6400 \text{ mol} + 0.4378 \text{ mol}}$$

$$= 0.5938$$

mole fraction of octane = X_o

$$= \frac{0.4378 \text{ mol}}{0.6400 \text{ mol} + 0.4378 \text{ mol}}$$

$$= 0.4062$$

Finally, you are ready to determine the vapor

pressure of each component of the mixture.

P(benzene) $= X_b P^o = 0.5938$ x (95.2 mm Hg)

$= \underline{\text{56.5 mm Hg}}$

P(octane) $= X_o P^o = 0.407$ x (14.1 mm Hg)

$= \underline{\text{5.72 mm Hg}}$

5. Rearranging the defining equation for osmotic pressure

$$M = \frac{\Pi}{RT} = \frac{0.0675 \text{ atm}}{(0.08206 \text{ L·atm/mol·K})(298 \text{ K})}$$

$$M = 0.002760 \text{ mole/L}$$

Using the definition of molarity

$$\text{molarity} = \frac{\text{mol of solute}}{\text{Liters of solution}}$$

$$\text{molarity} = \frac{\text{g of solute/molar mass}}{\text{Liters of solution}}$$

$$\text{molar mass} = \frac{95.2 \text{ g}}{(0.00276 \text{ M})(1.95 \text{ L})}$$

$$\text{molar mass} = \underline{\text{17,700 g/mole}}$$

6. We must assume some quantity of solvent, solute, or solution, and as usual, it is simplest to assume some amount of the substance in the denominator of the definition of the concentration unit that is provided, that is, weight percent. Thus, we will assume 1000 grams of solution.

It is apparent then that the solution will consist of 300.0 grams of ethylene glycol and 700.0 grams of water.

$$\text{moles of solute} = \frac{\text{grams of solute}}{\text{molar mass}}$$

$$\text{moles of solute} = \frac{300.0 \text{ g}}{62.07 \text{ g/mol}} = 4.833 \text{ mole}$$

The kilograms of solvent is 0.7000 kg, and these values can be substituted into the defining equation for molality.

$$\text{molality} = \frac{\text{moles of solute}}{\text{kilograms of solvent}}$$

$$\text{molality} = \frac{4.833 \text{ moles}}{0.7000 \text{ kg}} = 6.905 \text{ m}$$

Finally, this value can be substituted into the freezing point depression relationship

$$\Delta T = K_f m = -1.86 \text{ }^\circ C/m \times 6.905 \text{ m}$$

$$\underline{\Delta T = -12.8^\circ C}$$

Freezing point of the radiator solution = $-12.8^\circ C$.

7. a. An emulsion is a colloidal dispersion of one liquid in another.
b. A suspension is <u>not</u> a colloidal dispersion, since the particles are so large that it will settle out fairly rapidly.
c. A sol is a general term for a colloidal dispersion. The term may be modified by addition of a prefix, for example, an aerosol is a colloid that is dispersed in a gas.
d. A gel is a colloidal dispersion that has solidified and so resists flow.

8. Using Henry's law

$$\text{molality} = kP$$

rearranging

$$P(\text{mm Hg}) = \frac{m}{k} = \frac{0.050 \ m}{2.6 \times 10^{-4} \ m/\text{mm Hg}}$$

$P(\text{mm Hg}) = 192 \ \text{mm Hg}$

$P(\text{atm}) = 192 \text{mm Hg}/760 \ \text{mm/atm}$

$\underline{P(\text{atm}) = 0.25 \ \text{atm}}$

PRACTICE TEST (45 min)

1. Calculate the osmotic pressure of a solution prepared by dissolving 42.8 grams of sucrose (molar mass = 342.3 g/mole) in 250.0 milliliters of water at a temperature of $17.0^{\circ}C$.

2. An aqueous solution contains 147.2 grams of H_2SO_4 in exactly 1.00 liter of solution. The density of the solution is 1.090 g/mL. Calculate the (a) molarity, (b) molality, (c) weight percent, and (d) mole fraction of H_2SO_4 in this solution.

3. The freezing point depression constant for benzene is $-4.90^{\circ}C/m$, and the freezing point of pure benzene is $5.5^{\circ}C$. What is the freezing point of the solution that results when 8.15 grams of the organic compound having the formula $C_2H_2Cl_4$ is dissolved in 151 grams of benzene? Assume that the organic compound doesn't dissociate in benzene.

4. Hexane is an organic compound having a molar mass of 86.2 g/mole and a vapor pressure of 153 mm Hg at $25^{\circ}C$. When 25.0 grams of an unknown organic liquid is mixed with 50.0 grams of hexane, the vapor pressure of hexane over the resulting solution is lowered to 107 mm Hg. Assuming ideal behavior, what is the molar mass of the unknown liquid?

5. Carbonated soft drinks are produced by saturating water with carbon dioxide under pressure. In one such operation, the pressure of the carbon dioxide is 3.5 atmospheres at $25^{\circ}C$. What is the concentration of the CO_2 in the soft drink

under these conditions? (At 25°C the Henry's law constant = 4.44x10^{-5} m/mm Hg for CO2.)

6. What is the boiling point of a solution prepared by dissolving 120.0 grams of NaCl in 500.0 grams of water? You may assume that the NaCl is 100% dissociated. (K_b = 0.52°C/m for water)

7. Calculate the vapor pressure of water over a 0.50 m aqueous solution of K_2SO_4 at 50°C. The vapor pressure of water at 50°C is 92.5 mm Hg. You may assume the potassium sulfate is 100% dissociated.

COMMENTS ON ARMCHAIR EXERCISES

1. What effect will the high salt concentration have on the vapor pressure of that solution compared to the pure water? Would you expect the two solutions to display a different tendency for water molecules to escape from the surface of the water?

2. Think about why the oil and water don't mix. Is one polar and one nonpolar? Which is which? Remember how the detergents acts as a bridge to pull grease and stains into water in the washing machine. Will it do the same thing here?

3. This is a simple application of osmosis. The potato consists mainly of water, and the amount of water can be increased or decreased, depending on the external solute concentration.

CONCEPT TEST ANSWERS

1. solution
3. molality
5. weight percent
7. supersaturated
9. polar, nonpolar
11. decreases
13. mole fraction
15. decrease further
17. solution
19. water

2. colligative
4. one
6. solubility
8. miscible
10. partial pressure
12. endothermic
14. higher, lower
16. lower
18. emulsions
20. hard

PRACTICE TEST ANSWERS

1. **(L5)** 11.9 atm
2. **(L1)** a. 1.50 M   b. 1.59 m   c. 13.5%
   d. mole fraction = 0.0278
3. **(L4)** 3.92°C
4. **(L3)** 1.00×10^2 g/mol
5. **(L2)** molality CO_2 = 0.118 m
6. **(L4)** 104.3°C
7. **(L3)** mole fraction of water = 0.974,
   vapor pressure = 90.1 mm Hg

CHAPTER 15
PRINCIPLES OF CHEMICAL
REACTIVITY: CHEMICAL KINETICS

CHAPTER OVERVIEW

Two basic factors control whether or not reactions will occur. The first of these, thermodynamics, has already been discussed, and the book will return to this topic again later on. As you may have already noted, even though a reaction appears to be spontaneous based on thermodynamic arguments, it still may occur so slowly that the speed with which the reaction progresses is negligible. This demonstrates the importance of the second factor, kinetics.

Just because the energy of the products is less than the energy of the reactants, it is not obvious that a reaction must occur. For example, consider all of the items around you that could burn in the air, but they don't catch fire unless some energy is first added. In order to fully understand the nature of a chemical reaction, it is necessary to determine the individual steps which combine to cause the reaction to occur, that is, the reaction mechanism. A key component in identifying the mechanism is the determination of the rate equation, and so this process is a major focus of this chapter.

Catalysts can accelerate reactions that would otherwise be slow, and they are very common both in the chemical industry as well as in nature.

LEARNING GOALS

1. One of the fundamental pieces of information for kinetic studies is the rate of a chemical reaction. It's important to be aware of the factors that affect the speed of a chemical process, know something about how reaction rates are measured, and be able to distinguish between initial rate and instantaneous rate. (Sec. 15.1)

2. One way to express the effects of reactant concentrations on the rate of a reaction is by means of a rate expression. Know how to determine the rate expression from data that shows how the initial rate varies with changes in the concentration of one of the reactants (Sec. 15.3)

3. It's also important to understand how the integrated form of the first order rate expression and the half-life of a reaction can be a measure of reaction rate. Be able to use the first order equation

$$\ln \frac{[A]}{[A]_o} = -kt$$

to perform rate calculations on simple systems and know how to determine the rate expression by graphing the time against the natural logarithm of the concentration and the reciprocal of the concentration to determine whether the reaction is first or second order. (Sec. 15.4)

4. The transition state theory and the collision theory provide an explanation of how chemical reactions occur. Understanding these theories is an important theoretical background for all of the discussions in this chapter. (Sec. 15.5)

5. Be able to use the Arrhenius equation

$$\ln \frac{k_2}{k_1} = - \frac{E^*}{R} \left[\frac{1}{T_2} - \frac{1}{T_1} \right]$$

to predict the effects of temperature on reaction rate. This will require that the equation be solved either directly or by means of a graph. (Sec. 15.5)

6. Understand how elementary steps are combined to develop a reaction mechanism and the meaning of the terms reaction intermediate and rate determining step. Although students working with kinetics for the first time may not yet have enough experience to propose reaction mechanisms, it should be possible to examine a proposed mechanism and determine if it agrees with the available experimental data. You should also understand how a catalyst may act to change the mechanism and rate

of a reaction. (Sec. 15.6 and 15.7)

ARMCHAIR EXERCISE

1. One of the interesting results from reaction rate studies is the observation that when water containing the heavier isotope, deuterium, is used in a reaction instead of normal water, the normal water reacts faster, despite the fact that both should be chemically identical. Can you use the kinetic molecular theory to explain this observation?

2. You arrive at a picnic, and discover that the beverages are still warm because someone forgot to put them in the refrigerator. As you wait for the beverages to cool, a discussion develops regarding how long it will take for the drinks to reach the ideal drinkable temperature. Someone suggests that it would be easy to determine; just measure how much the temperature falls during the first five minutes the cans are in the refrigerator, then use a simple ratio. You remember that this kind of cooling is basically a first order process and disagree. What is the basis for your argument?

CONCEPT TEST

1. The _____ rate of a chemical

reaction is determined by drawing a line tangent to

the concentration-time curve at a particular time

and obtaining the slope of this line.

2. For a rate equation of the form

   $$Rate = k[A]^2$$

k, the proportionality constant, is known as the

_____.

3. The relationship between reactant

concentration(s) and reaction rate is expressed by an equation called the _____.

4. If for the equation

$$2 A + B \rightarrow A_2B$$

the rate expression is

$$Rate = k [A][B]$$

what is the order of the reaction? _____

5. How is the relation between rate and concentration of a reactant determined?

a. experimentally   b. from the balanced equation

c. neither of these

6. The sum of all the exponents in the rate equation is called the _____.

7. The half-life of a certain first-order reaction is 2.2 hours. What fraction of the original reactant in this process will remain after 8.8 hours? _____

8. What is the value of the rate constant for a reaction that has a half-life of 2.2 hours?

_____

9. For a first order reaction, the slope of the line that results when ln[A] is plotted against t is equal to _____.

10. For a first order reaction, you will obtain a straight line if you plot _____ against time; for a second order reaction, you must plot _____ against time to obtain a straight line.

11. In order to react, molecules must have an energy in excess of the _____ for the reaction.

12. According to the Arrhenius equation, we should obtain a straight line if we plot ln k against _____.

13. The _____ is the collection of bond-breaking and bond-making steps that occurs during the conversion of reactants to products.

14. It is a good rule of thumb that reaction rates will double for every _____ degree Celsius raise in temperature.

15. _____ is the molecularity of the elementary process

$$2 A + B \rightarrow A_2B.$$

How probable is it that this elementary process will occur? _____.

16. The molecularity and the _____ of an elementary step are always the same.

17. The rate of the overall reaction is limited by and is exactly equal to the combined rates of all elementary steps up through the _____ step in the mechanism.

18. A heterogeneous catalyst is one that is present in a different _____ from the reactants being catalyzed.

19. A catalyst accelerates a reaction by altering the mechanism so that the _____ is lowered.

20. A reagent that accelerates a chemical reaction but is not, itself, transformed in the process, is called a(n) _____.

21. When a change in the concentration of a reactant has no effect on the rate of the reaction, the order of that reactant in the rate expression is _____.

STUDY HINTS

1. Although the exponents in the rate expression will only be the same as the stoichiometric coefficients in the balanced equation by coincidence, it's important to remember that the exponents in the rate expression for an elementary process are always equal to the coefficients of the reactants in that elementary process.

2. There are three basic types of problems that involve the determination of the overall rate of a

chemical reaction. Although these situations are quite different, students will sometimes confuse them. In the first case, you will be given <u>initial</u> rates and reactant concentrations from <u>several different experiments</u>; this type of problem is best solved by inspection. In the second type, you are told the concentration of a reactant at <u>a number of different times</u>; a graphical solution is usually best here. A third case involves using a proposed mechanism to predict a rate expression. When you encounter one of these problems, look carefully at what data is given, and that will usually lead you to the correct method of solution.

Don't confuse the determination of reaction order with other problem types. The most common source of confusion is the use of the Arrhenius equation to determine the variation of reaction rate with temperature. Observe carefully how these problem types differ, and it should be easy to recognize them as different cases.

3. Read carefully when doing problems that use the integrated rate equation. The information provided may be either the amount that has reacted or the amount that remains. Remember that the integrated rate equation requires the concentration at some specific time, indicated by the value of t. Unless you are cautious, it's easy to confuse the amount that has reacted with the amount that didn't react. This is a common source of errors in this problem type. Practice problems 4b and 4c are examples of this situation.

4. When using a proposed reaction mechanism to predict a rate expression, remember that this prediction is only a theory. If it disagrees with the experimental rate expression, it is necessary to revise the mechanism. On the other hand, the fact that a certain mechanism does agree with the experimental rate expression doesn't insure that the mechanism is correct. For many years the reaction

$$2 \ HI(g) \rightarrow H_2(g) + I_2(g)$$

was given as a classic example of a bimolecular reaction, but it was finally discovered that the mechanism was much more complicated (<u>J. Chem.</u>

<u>Phys.</u>, **36**, 1925, (1962)).

5. Remember that mathematically

$$\ln A/B = \ln A - \ln B$$

In some problems you will need to use this relationship to solve for either the initial concentration or the concentration at the given time.

6. Remember that the R value in the Arrhenius equation has different units (and value) from that which you have used with the gas laws.

PRACTICE PROBLEMS

1. **(L2)** On the left of the table on the next page are various possible changes in the concentration of a reactant, A. Fill in the blanks by determining the effect of the change indicated on a reaction that has the rate expression given at the top of the column.

Change in [A]   Rate = [A]   Rate = [A]^2

double [A]   _____   _____

triple [A]   _____   _____

halve [A]   _____   _____

one-fourth [A]_____   _____

quadruple [A] _____   _____

2. **(L2)** The data below was collected for the hydrolysis of a simple sugar in aqueous solution at 23 degrees Celsius.

[sugar]	time
2.00 mmol/dm^3	0 min
1.62	60
1.31	120
0.861	240
0.565	360

Determine whether this is a first or second order reaction by obtaining a separate sheet of graph paper and graphing (a) ln [sugar] vs. time and then (b) 1/[sugar] vs. time. In which case do you obtain a straight line? What is the order of the reaction? From the slope of the line, what is the value of k for this reaction?

3. **(L2)** The decomposition of gaseous hydrogen iodide at $716°C$ is described by the equation

$$HI_{(g)} \rightarrow 1/2 \ H_{2(g)} + 1/2 \ I_{2(g)}$$

The results of several different rate experiments are shown in the table below.

Experiment Number	Initial [HI] (mole/liter)	Initial Rate (mole/liter/min)
1	4.0×10^{-3}	1.07×10^{-9}
2	6.0×10^{-3}	2.41×10^{-9}
3	8.0×10^{-3}	4.29×10^{-9}

a. Write the rate expression for this reaction.

b. What is the overall order of this reaction?

c. Calculate the value of the rate constant for this reaction.

4. **(L3)** a. The rate constant is 2.84×10^{-5} yr^{-1} for the first-order radioactive decay of plutonium-239. a. What is the half-life for this isotope? b. How long must a given sample of this isotope remain undisturbed until the amount of plutonium remaining is 20.0% as great as the original amount?

c. Based on the information provided in part a, how long must a sample of plutonium-239 remain undisturbed in order to allow 20.0% of the original plutonium to undergo radioactive decay?

5. **(L6)** At $320°C$ the gas phase reaction

$$SO_2Cl_2 \quad \rightarrow \quad SO_2 \quad + \quad Cl_2$$

is first order with a rate constant of 2.0×10^{-5} s^{-1}. Which of the following mechanisms agrees with this observation?

Mechanism I: $SO_2Cl_2 \quad \rightarrow \quad SO_2 \quad + \quad Cl_2$
(simple, elementary step)

Mechanism II: $2\ SO_2Cl_2 \quad \rightarrow \quad S_2O_4Cl_4 \quad$ (slow)

$\qquad\qquad\qquad S_2O_4Cl_4 \quad \rightarrow 2\ SO_2 \quad + 2\ Cl_2 \quad$ (fast)

6. **(L5)** Determine the activation energy (E_a) for the reaction

$$N_2O_{5(g)} \quad \rightarrow \quad 2\ NO_{2(g)} \quad + \quad 1/2\ O_{2(g)}$$

based on the following observed rate constants and temperatures:

Temperature (°C)	k (sec^{-1})
25	3.46×10^{-5}
55	1.50×10^{-3}

7. **(L6)** Can you suggest an explanation of why termolecular gas reactions are relatively unlikely? Under what conditions would you expect termolecular reactions to be most probable?

8. **(L2)** The elimination of methylmercury, a toxic mercury compound, from the human body is a first order process with an estimated half-life of 70 days. Individuals who eat large amounts of fish contaminated with methylmercury may have blood concentration of this compound as high as 0.200 mg/liter. If a person with this blood level of mercury stopped all mercury intake, calculate the concentration of methylmercury that would remain in the blood after 350 days.

PRACTICE PROBLEM SOLUTIONS

1.

Change in [A]	Rate = [A]	Rate = $[A]^2$
2x[A]	rate doubles	rate quadruples
3x[A]	rate triples	rate 9 times greater
1/2 [A]	rate 1/2 as great	rate 1/4 as great
1/4 [A]	rate 1/4 as great	rate 1/16 as great
4x[A]	rate 4 times as great	rate 16 times as great

2. Based on the graphs below the reaction must be first order since only the ln[sugar] vs t graph gave a straight line.

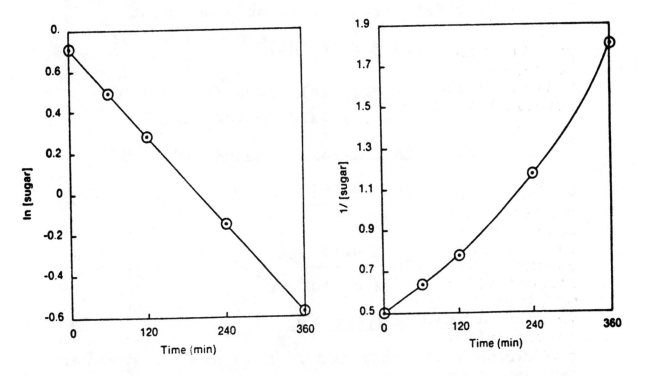

Figure 15.1 Graphs of ln[sugar] vs. t and 1/[sugar] vs. t for the hydrolysis of a sugar solution

2. (cont.) Now determine the slope of the above

$$k = \frac{(\ln[sugar]_5 - \ln[sugar]_1)}{(t_5 - t_1)} = \frac{(-0.5674 - 0.69314)}{(360 - 0)}$$

275

$k = 0.00355$

3. a. Compare the first and third experiments. The initial concentration of the hydrogen iodide doubles from experiment 1 to experiment 3; however, the rate becomes four times as great. This indicates a second order relationship, and so the rate equation must be

$$Rate = k[HI]^2$$

b. There is only one reactant, and so the overall order is second order.

c. Using the data from the first experiment

$$k = Rate/[HI]^2$$

$$k = \frac{1.07 \times 10^{-9} \text{ mole/liter/min}}{(4.0 \times 10^{-3} \text{ mole/liter})^2}$$

$$k = 6.7 \times 10^{-5} \text{ (mole/liter)}^{-1}/min$$

4. a.

$$t_{1/2} = \frac{0.693}{k}$$

$$t_{1/2} = \frac{0.693}{2.84 \times 10^{-5} \text{ yr}^{-1}}$$

$$t_{1/2} = 2.44 \times 10^4 \text{yr}$$

4.b. Write the integrated form of the first-order rate equation,

$$\ln \frac{[A]}{[A_o]} = -kt$$

Notice that $[A] = 0.200 [A_o]$, and substitute the values provided

$$\ln \frac{0.200 [A_o]}{[A_o]} = -(2.84 \times 10^{-5} \text{ yr}^{-1})t$$

276

$$\underline{t = 56,700 \text{ year}}$$

4. c. Begin again with the integrated first-order rate equation,

$$\ln \frac{[A]}{[A_o]} = -kt$$

But this time $[A] = [A_o] - 0.200\,[A_o] = 0.800[A_o]$. Otherwise the values are the same.

$$\ln \frac{0.800\,[A_o]}{[A_o]} = -(2.84\times10^{-5}\text{ yr}^{-1})t$$

$$\underline{t = 7,860 \text{ years}}$$

Notice how a small change in the wording caused a major change in the answer.

5. In the first mechanism, the rate is controlled by a unimolecular process, and so the reaction should be first order. This agrees with observation, so this mechanism may be correct.

   If the second proposed mechanism is correct, the first step controls the rate, and since that step is bimolecular, the order of that step and the overall reaction must be second order. This doesn't agree with the experimental observation, and so that mechanism cannot be correct.

6. Using the Arrhenius equation

$$\ln \frac{k_2}{k_1} = - \frac{E^*}{R}\left[\frac{1}{T_2} - \frac{1}{T_1} \right]$$

Substituting the values given

$$\ln \frac{1.50\times10^{-3}}{3.46\times10^{-5}} = - \frac{E^*}{8.314 \text{ J/mol·K}}\left[\frac{1}{328K} - \frac{1}{298K} \right]$$

$$3.7694 = \frac{E^*}{8.314 \text{ J/mol·K}}(3.0692\times10^{-4})$$

$E_a = \underline{1.02 \times 10^5 \text{ J/mol}}$ or 102 kJ/mole

7. Remember that the molecules of a gas occupy only a small fraction of the total volume of the gas sample. The large volume of empty space makes it unlikely that three molecules will simultaneously collide. In order to make these reactions more likely it is necessary to increase the number of collisions. This could be accomplished by increasing the temperature, so that the molecules move faster, or by decreasing the volume, so that there is less empty space and greater probability of collisions.

8. The elimination rate constant, k_e, can be obtained from the half-life using the equation

$$k_e = \frac{0.693}{t_{1/2}} = \frac{0.693}{70 \text{ days}} = 0.0099 \text{ days}^{-1}$$

This allows you to solve the integrated first-order rate equation

$$\ln \frac{[A]}{[A_o]} = -kt$$

Substituting the values provided

$$\ln \frac{[A]}{0.200 \text{ mg/liter}} = -(0.0099 \text{ day}^{-1})(350 \text{ days})$$

$$\ln[A] = -(0.0099 \text{ day}^{-1})(350 \text{ days}) + \ln(0.200 \text{ mg/liter})$$

$$\underline{[A] = 0.006 \text{ mg/liter}}$$

PRACTICE TEST (50 Minutes)

1. What four conditions are generally considered to determine the rate of a chemical reaction?

2. At moderate conditions of temperature and pressure, each of the following reactions occurs by means of a single-step, elementary process. Write the rate expression for each of these examples, and

give the order and molecularity of each.

a. $OH^- + CH_3Br \rightarrow CH_3OH + Br^-$
b. $Br(g) + H_2(g) \rightarrow HBr(g) + H(g)$

3. For the hypothetical reaction
$$A + B = C$$
The activation energy for the combination of A and B to form C is 45.0 kJ/mole. The reverse process, C forming A and B, has an activation energy of 32 kJ/mole. Is the reaction exothermic or endothermic as written above?

4. If a human ingests dichlorophenoxyacetic acid, a commonly used herbicide, the elimination in the urine can be approximated as a first-order process having a half-life of 220 hours. Suppose that a person accidentally swallows some of this compound. How long will be required before the amount in the body is only 20.0% as great as the amount ingested?

5. A plot of ln K against 1/T for a certain reaction gives a straight line with a slope equal to -5300 mole K and an intercept of -24.7. Calculate the activation energy, E_a, for this reaction.

6. Photochemical smog, a type of air pollution common in cities, is partially caused by reactions between ozone and hydrocarbons (emitted from automobiles). One example of this type of reaction is the combination of ozone and ethylene

$$C_2H_{4(g)} + O_{3(g)} \rightarrow H_2CO \text{ (formaldehyde)} + H_2CO_{2(g)}$$

A kinetic study has produced the following data.

Experiment	Initial $[O_3]$	Initial $[C_2H_4]$	Initial rate
1	1.6×10^{-9} M	1.0×10^{-9} M	1.94×10^{-28} mol/L/min
2	4.0×10^{-9}	1.0×10^{-9}	4.84×10^{-28}
3	1.6×10^{-9}	3.0×10^{-9}	5.81×10^{-28}

a. Write the rate expression for this reaction.

b. What is the overall order of this reaction?

c. Calculate the initial rate of this reaction if both O_3 and C_2H_4 are initially 2.0×10^{-9} M

7. For reasons best left to the imagination, certain insects, such as the common firefly, display their presence by a flashing light that is produced by a complex set of biochemical reactions. The rate at which fireflies flash varies with the temperature, since the rate-controlling step in this process is temperature dependent. It is observed that the time between flashes for a typical firefly is 16.3 seconds at 21°C, and the time between flashes is 13.0 seconds at 27.8°C. Based on this information, calculate the activation energy for the chemical reaction that controls the flashing. (Hint: Notice that you are given the time between flashes, which is inversely related to the rate of flashing.)

8. The rate expression for the reaction

$$2\ NO(g)\ +\ 2\ H_2(g)\ \rightarrow\ N_2(g)\ +\ 2\ H_2O(g)$$

has the form

$$Rate\ =\ k[NO]^2[H_2]$$

Consider the following two mechanisms that have been suggested for this process:

MECHANISM I

$$NO\ +\ H_2\ \rightarrow\ H_2O\ +\ N\quad (slow)$$

$$N\ +\ H_2\ +\ NO\ \rightarrow\ N_2\ +\ H_2O\ (fast)$$

MECHANISM II

$$2\ NO\ +\ H_2\ \rightarrow\ N_2\ +\ H_2O_2\quad (slow)$$

$$H_2O_2\ +\ H_2\ \rightarrow\ 2\ H_2O\qquad\qquad (fast)$$

Based on the available information, which of these mechanisms is possible?

9. At 1000°C, cyclopropane, an organic compound, reacts according to the equation

$$
\begin{array}{ccc}
& CH_2 & \\
& /\backslash & \\
H_2C & - \ CH_2 &
\end{array}
\quad \rightarrow \quad
H_2C = \overset{\overset{\textstyle H}{|}}{C} - CH_3
$$

This is a first order reaction with a half-life of 7.5×10^{-2} seconds. Calculate the length of time necessary for 90.0% of some original amount of cyclopropane to react under these conditions.

COMMENTS ON ARMCHAIR EXERCISES

1. Although both hydrogen and deuterium are the same chemically, the mass difference can cause these two atoms to react differently. Consider what you know about the relationship between molecular velocities and mass. Does this difference seem likely to affect the ability to react?

2. The suggested way of determining the cooling time assumes that there is a linear relationship between temperature and time. If this is a first order process, is that assumption accurate?

CONCEPT TEST ANSWERS

1. instantaneous rate
2. rate constant
3. rate equation or rate law
4. second
5. a. experimentally
6. overall order
7. one-sixteenth
8. 0.32 hr^{-1}
9. -k (or -rate constant)
10. ln [A], 1/[A]
11. activation energy
12. 1/T
13. reaction mechanism
14. 10°C
15. termolecular, not very
16. order
17. slowest
18. phase
19. activation energy
20. catalyst
21. zero

PRACTICE TEST ANSWERS

1. (L1) concentration of reactants
   temperature
   presence of a catalyst
   physical state of the reactants

2. **(L6)**
a. Rate = $- k[OH^-][CH_3Br]$, bimolecular, second order

b. Rate = $- k[Br][H_2]$, bimolecular, second order

3. **(L6)** The reaction is endothermic, with an enthalpy of reaction of 13 kJ/mole
4. **(L3)** 510 hours
5. **(L5)** 44 kJ
6. **(L2)** a. Rate = $k[O_3][C_2H_4]$

   b. Second order

   c. 4.8×10^{-28} mol/L/min
7. **(L5)** 24.5 kJ
8. **(L7)** In mechanism I, the first step is the rate controlling reaction, and based on this equation, the rate expression should be first order in NO and first order in H_2. This is contrary to observation, and so must be wrong.

   In mechanism II, the first step is also the rate controlling reaction, and this equation produces a rate expression that would be second order in NO and first order in H_2. This agrees with observation, and may be correct.
9. **(L3)** 0.25 sec

CHAPTER 16
PRINCIPLES OF REACTIVITY: CHEMICAL EQUILIBRIUM

CHAPTER OVERVIEW

Most of the reactions discussed thus far appear to have gone to completion, that is, the reactants were converted into products to the extent possible based on the stoichiometry of the reaction. Actually most chemical reactions may, under the proper circumstances, reach a dynamic equilibrium. The forward reaction is opposed by a reverse reaction (the products acting as the reactants) which is proceeding at exactly the same rate. This situation is important, since in nature many reactions occur under conditions that cause this equilibrium to favor either reactants or products. In addition, many industrial processes are based on manipulating chemical equilibria to produce substances that are desirable or prevent the production of those that are undesirable.

This chapter discusses many different ways to perform quantitative determinations on equilibrium systems. It also outlines a qualitative method, based on the ideas of le Chatelier, that can be particularly useful for estimating how an equilibrium may shift due to changes in conditions. Notice, however, that all of these discussions assume that equilibrium is actually present. Your first step should always be to ask if the system is truly in equilibrium to insure that the methods you will use are valid.

LEARNING GOALS

1. The ability to write correct equilibrium expressions for both homogeneous and heterogeneous equilibria is essential. (Sec. 16.2)

2. Equilibrium constants for gas reactions may be stated in terms of either concentrations or pressures. Be able to convert either from K_c to K_p

or from K_p to K_c and to obtain the appropriate form of the equilibrium constant for a given situation. When given the equilibrium constant for one balanced equation, know how to obtain the new equilibrium constant if the original equation is reversed, multiplied by a constant, or combined with other balanced equations. (Sec. 16.2)

3. It's important to understand the reaction quotient, since it is valuable determining whether or not equilibrium conditions exist and the direction a reaction will shift to attain equilibrium. (Sec. 16.3)

4. The value of K, the equilibrium constant, can be calculated from experimental data on the percent of reaction, the concentration of the reactant that actually undergoes the process, or the equilibrium concentrations of each species in the system. Know how to calculate K in each case. (Sec. 16.4)

5. If you are given the value of the equilibrium constant, there are several different problem types that require you to calculate relationships between reactant and product concentrations. Be familiar with these various problem types. (Sec. 16.5)

6. Be able to use le Chatelier's principle to predict how a given equilibrium system will be affected by changes in temperature, volume, or in concentrations of the components. (Sec. 16.6)

7. Where appropriate, be able to use the relationship between reaction mechanism and equilibrium to obtain the rate expression from a mechanism having a fast equilibrium as the first step. (Sec. 16.7)

ARMCHAIR EXERCISES

1. Sometimes at novelty shops you will see an inexpensive "weather predictor", that is supposed to turn color from blue to pink when it is going to rain. Actually the device consists of some object that has been coated with cobalt chloride. How does this device work, and do you think that it will make accurate predictions? HINT: Cobalt chloride will undergo the following equilibrium:

$$[Co(H_2O)_6]Cl_2(s) \rightleftharpoons [Co(H_2O)_4]Cl_2(s) + 2 H_2O(g)$$

2. At your local toy store, you should be able to find toy cars that will change color when placed in hot water. For example, one such car changes from red to pink when heated gently. If the pink car is placed in an ice bath, it will turn red again. The chemistry of the change is too complicated for this course, but based on what you know, you should be able to determine whether the conversion from red to pink is an exothermic or an endothermic reaction.

3. If you buy unripe fruit or vegetables at the store, you run the risk that the produce may spoil before it becomes ripe enough to eat. The ripening process is caused by ethene gas (commonly known as ethylene) which is naturally produced by tomatoes, apples, peaches, and other fruits and vegetables. Can you think of a way that you can take advantage of this information to help your produce ripen faster?

CONCEPT TEST

1. The _____ relates concentrations of reactants and products at equilibrium at a given temperature to a numerical constant.

2. When a chemical system is at equilibrium, the forward and reverse reactions continue, but they take place at _____.

3. In the equation

$$K_p = K_c(RT)^{\Delta n}$$

the term $\Delta n =$ _____.

4. The concentrations of what types of substances never appear in the equilibrium constant expression?

a. _____

b. _____

c. _____

5. When all of the stoichiometric coefficients in a balanced chemical equation are multiplied by a constant factor, the equilibrium constant for the old equation must be _____ to obtain the new equilibrium constant.

6. When a balanced chemical equation is written in reverse, that is, the old reactants become the new products and the old products become the new reactants, the equilibrium constant for the old equation is the _____ of the new equilibrium constant.

7. When two or more balanced chemical equations are added to produce a new chemical equation, the equilibrium constant for the new equation will be equal to the _____ of the equilibrium constants for the reactions that were added.

8. A system is at equilibrium only when the

reaction quotient, Q, is equal to _____.

9. When _____, this means the reaction is product favored, that is, the equilibrium concentration of products are greater than the equilibrium concentrations of reactants. (Choose from a. K >> 1, b. K << 1, or c. K = 1)

10. The reaction quotient and the equilibrium constant expression for a given balanced equation look the same, but the difference between them is that _____.

11. Le Chatelier's principle states that a change in a system in chemical equilibrium will cause the system to change in such a manner as to _____ _____ the effect of the change.

STUDY HINTS

1. When students first begin to write equilibrium expressions, it is sometimes tempting to include the plus sign from the balanced equation, which is, of course, wrong. Thus for the equation

$$A + B \rightleftharpoons C$$

They might write an equilibrium expression in the <u>incorrect</u> form

$$K = \frac{[C]}{[A] + [B]} \quad \text{INCORRECT!}$$

The mistake comes from writing down the reactants without thinking. More experience (and loss of points on some tests) quickly cures this error, but it's better to be sure to avoid this mistake from the beginning.

2. Most students quickly recognize how to use Le Chatelier's principle to predict the effects of changing system conditions on a chemical equilibrium, but this type of question frequently contains tricks that will trap the unwary. Remember that pure solids, pure liquids, and solvents in dilute solutions don't appear in the equilibrium expression. This also means that adding more of these substances to (or removing them from) an equilibrium system will have no effect on the equilibrium.

Also notice that the shift in the equilibrium doesn't completely eliminate the original stress. For example, if I_2 is added to the system

$$2\ HI(g) \quad \rightleftarrows \quad H_2(g) \quad + \quad I_2(g)$$

the equilibrium does shift to the left, and the concentrations of hydrogen and iodine gases are less than they were <u>after the extra iodine was added</u>. It is important to realize that the final concentration of iodine gas is greater than it was <u>before</u> you added the extra iodine.

3. As the name implies, an equilibrium expression is not a true equality unless all of the concentration values used are equilibrium concentrations. As the problems become more complicated, keep in mind that the ultimate purpose of what we are doing is usually to obtain equilibrium values so that we can use them in the equilibrium expression.

4. Only concentrations or pressures can be used in the equilibrium expression. In some problems the number of moles and the volume of the container or the number of moles and the volume of the solution are given rather than concentrations. Don't forget to convert these values to concentration units. Whenever a problem statement provides the volume of the aqueous solution or the volume of a gas container, check to see if a conversion is necessary to obtain molar concentrations.

5. This is a good time to review the method for solving a quadratic equation, since this skill will

be necessary for some of the problems in this chapter.

PRACTICE PROBLEMS

1. **(L6)** The following mixture is at equilibrium in a closed container

$$Ti(s) + 2 Cl_2(g) \rightleftharpoons TiCl_4(g) + energy$$

Listed on the left below are several changes that may affect the equilibrium composition of this system. For each separate change indicated, place an X in the appropriate space on the right to indicate whether that single change will increase, decrease, or have no effect on the concentration of chlorine gas present in the system after equilibrium has been restored.

	increase $[Cl_2]$	decrease $[Cl_2]$	no effect
a. Add $TiCl_4$ to the system	X		
b. Add Ti to the system			X
c. Decrease the temperature		X	
d. Add an inert gas to increase the pressure	~~X~~	X	
e. Add Cl_2 to the system	X		

2. **(L1)** Write equilibrium expressions for each of the following reactions:

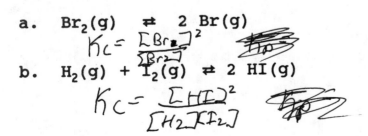

a. $Br_2(g) \rightleftharpoons 2 Br(g)$

$$K_c = \frac{[Br_\bullet]^2}{[Br_2]}$$

b. $H_2(g) + I_2(g) \rightleftharpoons 2 HI(g)$

$$K_c = \frac{[HI]^2}{[H_2][I_2]}$$

289

c. $2\ NOBr(g) \rightleftharpoons 2\ NO(g) + Br_2(g)$

$$K_c = \frac{[Br_2][NO]^2}{[NOBr]^2}$$

d. $TiCl_4(g) \rightleftharpoons Ti(s) + 2\ Cl_2(g)$

$$K_c = \frac{[Cl_2]^2}{[TiCl_4]}$$

e. $2\ HgO(s) \rightleftharpoons 2\ Hg(s) + O_2(g)$

$$K_c = [O_2]$$

f. $Ag^+(aq) + 2\ NH_3(aq) \rightleftharpoons Ag(NH_3)_2^+(aq)$

3. (L2) The value of K_p is 1.2 atm^{-2} at 377°C for the reaction

$$3\ H_2(g)\ +\ N_2(g)\ \rightleftharpoons\ 2\ NH_3(g)$$

Calculate the value of K_c for this reaction.

$$K_c = \frac{[NH_3]^2}{[H_2]^3[N_2]}$$

$$K_p = K_c RT^{\Delta n} \qquad K_p = 1.2\ atm^{-2}$$

$$\Delta n = moles\ produced - moles\ reacted$$

$$2 - 4 = -2$$

4. (L3) Silver chloride and silver iodide are both relatively insoluble in water. The equilibrium reactions and constants for the dissolving of these compounds are as follows:

$$AgCl(s) \rightleftharpoons Ag^+(aq) + Cl^-(aq) \qquad K = 1.8 \times 10^{-10}$$

$$AgI(s) \rightleftharpoons Ag^+(aq) + I^-(aq) \qquad K = 1.5 \times 10^{-16}$$

If solid AgCl and solid AgI are placed in water in separate beakers, which beaker will have the greater concentration of Ag$^+$?

AgCl

5. **(L4)** Calculate the value of the equilibrium constant, K_c, for the reaction

$$H_2(g) \quad + \quad Br_2(g) \quad \rightleftharpoons \quad 2\ HBr(g)$$

if the following equilibrium concentrations are observed at 1297 K, $[H_2]$ = $[Br_2]$ = 0.020 M and $[HBr]$ = 7.4 M

$$K_c = \frac{[HBr]^2}{[H_2][Br_2]}$$

$$K_c = \frac{7.4^2}{.02^2} = 1.36 \times 10^5$$

6. **(L4)** When 9.500 moles of PCl_5 gas is placed in a 2.00 liter container and allowed to come to equilibrium at a constant temperature, it is observed that 40.0% of the PCl_5 decomposes according to the equation

$$PCl_5(g) \quad \rightleftharpoons \quad PCl_3(g) \quad + \quad Cl_2(g)$$

What is the value of K_c for this reaction under these conditions?

7. **(L5)** When 1.00 mole of phosphorus pentachloride gas is placed in a 2.00 liter container at 250°C and allowed to react according to the equation

$$PCl_5(g) \quad \rightleftharpoons \quad PCl_3(g) \quad + \quad Cl_2(g) \qquad \text{initial M}$$

initial M   .5   0   0   M $\square$

$-x$ $+x$ $+x$

.5 $-x$ x x

291

the equilibrium constant for the process, K_c, has a value of 0.0415 M. Calculate the equilibrium concentration for each species in this system.

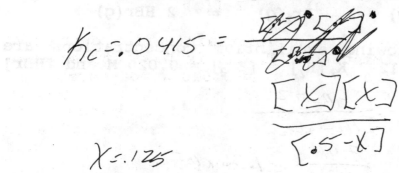

$K_c = .0415 =$

$$\frac{[X][X]}{[.5-X]}$$

$X = .125$

8. (L5) Calculate the equilibrium concentrations that would result in the previous problem if you initially added not only 1.00 mole of phosphorus pentachloride but also 0.200 moles of chlorine gas to the original system and allowed enough time to pass so that equilibrium conditions were restored.

9. (L5) At 400°C, the reaction below occurs when 1.50 moles of $POCl_3$ is placed in a 0.500 liter container

$$POCl_3(g) \rightleftharpoons POCl(g) + Cl_2(g) \qquad K_c = 0.248 \text{ M}$$

Calculate the number of moles of POCl that must be added to this system in order to produce an equilibrium concentration of Cl_2 equal to 0.500 M.

292

10. (L3) At 1000 K, $3.1 \times 10^{-3} = K_p$ for the reaction

$$I_2(g) \rightleftharpoons 2\ I(g)$$

It is observed that in a certain sealed container at 25°C the pressure of I_2 gas is 0.21 atmospheres and the pressure of I gas is 0.030 atmospheres. a. Is this system at equilibrium? If it is not at equilibrium, will the pressure of I_2 increase or decrease as it continues to approach equilibrium? b. If a catalyst is added to the system, what effect will it have on the equilibrium?

11. (L7) For the reaction

$$H_2(g)\ +\ I_2(g) \rightleftharpoons 2\ HI(g)$$

The rate expression is reaction rate = $k[H_2][I_2]$

Does the following proposed mechanism agree with the rate expression?

$$I_2 \underset{k_{-1}}{\overset{k_1}{\rightleftharpoons}} 2\ I \qquad \text{fast, equilibrium}$$

$$2\ I\ +\ H_2 \overset{k_2}{\rightarrow} 2\ HI \quad \text{slow, rate determining}$$

PRACTICE PROBLEM SOLUTIONS

1.

	increase $[Cl_2]$	decrease $[Cl_2]$	no effect
a. Add $TiCl_4$ to the system	X		
b. Add Ti to the system			X
c. Decrease the temperature		X	
d. Add an inert gas to increase the pressure		X	
e. Add Cl_2 to the system	X		

2. a. $K = \dfrac{[Br]^2}{[Br_2]}$ **b.** $K = \dfrac{[HI]^2}{[H_2][I_2]}$

c. $K = \dfrac{[NO]^2[Br_2]}{[NOBr]^2}$ **d.** $K = \dfrac{[Cl_2]^2}{[TiCl_4]}$

e. $K = [O_2]$ **f.** $K = \dfrac{[Ag(NH_3)_2^+]}{[Ag^+][NH_3]^2}$

3. The relationship between K_p and K_c is

$$K_p = K_c(RT)^{\Delta n}$$

Δn = moles of gas produced − moles of gas reacted
$= 2 - 4 = - 2$

$K_c = (1.2 \text{ atm}^{-2})(0.08206 \text{ L·atm/mole·K} \times 650 \text{ K})^2$

$\underline{K_c = 3.4 \times 10^3}$

4. The product of the ion concentrations is much smaller for AgI than for AgCl. Since both compounds produce the same number of moles of ions per mole in solution, this suggests that the individual ion values will be larger for AgCl, and so the silver ion concentration will be larger in the beaker containing the silver chloride.

5. A data table is useful here, even though the only values given are equilibrium concentrations, and so the data table has only one line.

	$[H_2]$	$[Br_2]$	$[HBr]$
equilibrium concentrations	0.020	0.020	7.4

Write the equilibrium expression

$$K_c = \frac{[HBr]^2}{[H_2][Br_2]}$$

Substitute the values from the equilibrium line of the data table

$$K_c = \frac{(7.4)^2}{(0.020)(0.020)}$$

$$\underline{K_c = 1.4 \times 10^5}$$

6. The first thing that should catch your attention is the fact that the volume of the container is provided. With this clue, it is easy to see that you are given moles of gas, not concentration of gas. Begin by calculating the initial concentration of PCl_5.

$[PCl_5]$ = 9.500 moles/2.00 liters = 4.750 M

If 40.0% of this gas reacts,
        the concentration reacting = 0.400 x 4.750 M
                                   = 1.896 M

You are now ready to write the data table for the reaction.

Data Table

	$[PCl_5]$	$[PCl_3]$	$[Cl_2]$
Before dissociation	4.750	0	0
Change	-1.896	+1.896	+1.896
Equilibrium conc.	2.854	1.896	1.896

Write the equilibrium expression

$$K_c = \frac{[PCl_3][Cl_2]}{[PCl_5]}$$

Substitute the values from the equilibrium concentration line of the data table

$$K_c = \frac{(1.896)(1.896)}{(2.854)}$$

$$\underline{K_c = 1.26}$$

7. a. Begin by calculating the initial concentration of the PCl_5.

initial $[PCl_5]$ = 1.00 mole/2.00 liter = 0.500 M

Now set up the data table, letting X = $[PCl_5]$ that reacts

	$[PCl_5]$	$[PCl_3]$	$[Cl_2]$
Initial conc.	0.500	0	0
Change	-X	+X	+X
Equilibrium conc.	0.500-X	X	X

Now write the equilibrium expression

$$K_c = \frac{[PCl_3][Cl_2]}{[PCl_5]}$$

Substitute the equilibrium concentrations from the data table.

$$0.0415 \text{ M} = \frac{(X)(X)}{(0.500-X)}$$

Cross multiply and rearrange

$$X^2 + 0.0415X - 0.02075 = 0$$

Compare this with the general quadratic form

$$aX^2 + bX + c = 0$$

and substitute the values for a, b, and c into the equation

$$X = \frac{-b + \sqrt{b^2 - 4ac}}{2a}$$

$$X = \frac{-0.0415 + \sqrt{(0.0415)^2 - 4(-0.02075)}}{2}$$

$$X = 0.125$$

Thus from the data table,

$$[PCl_5] = 0.500 - 0.125 = \underline{0.375 \text{ M}}$$

$$[PCl_3] = [Cl_2] = \underline{0.125 \text{ M}}$$

8. The initial concentration of the PCl_5 is still the same, and

initial $[Cl_2]$ = 0.200 mole/2.00 liter = 0.100 M

Now set up the data table, letting X = $[PCl_5]$ that reacts

	$[PCl_5]$	$[PCl_3]$	$[Cl_2]$
Initial conc.	0.500	0	0.100
change	-X	+X	+X
Equilibrium conc.	0.500-X	X	0.100+X

Now write the equilibrium expression

$$K_c = \frac{[PCl_3][Cl_2]}{[PCl_5]}$$

Substitute the equilibrium concentrations from the data table.

$$0.0415 \text{ M} = \frac{(X)(0.100+X)}{(0.500-X)}$$

Cross multiply and rearrange

$$X^2 + 0.1415X - 0.02075 = 0$$

Compare this with the general quadratic form

$$aX^2 + bX + c = 0$$

and substitute in the quadratic equation as in the previous problem

$$X = \frac{-0.1415 \pm \sqrt{(0.1415)^2 - 4(-0.02075)}}{2}$$

$$X = 0.0897$$

$[PCl_5] = 0.500 - 0.0897 = \underline{\text{0.410 M}}$

$[PCl_3]$ $\underline{\text{0.0897 M}}$

$[Cl_2] = 0.100 + 0.0897 = \underline{\text{0.1897 M}}$

Notice that the effect has been to shift the equilibrium to the left, exactly as Le Chatelier's principle would predict.

9. Calculate the initial concentration of $POCl_3$

$$[POCl_3] = 1.50 \text{ moles}/0.500 \text{ liters} = 3.00 \text{ M}$$

Now write the data table, letting X = [POCl] added

	$[POCl_3]$	$[POCl]$	$[Cl_2]$
Original conc.	3.00	X	0
Change	-0.500	+0.500	+0.500
Equilibrium conc.	3.00-0.500 =2.50	X+0.500	0.500

Notice that because the equilibrium concentration of chlorine was known to be 0.500 M, this required the reaction of 0.500 M of the $POCl_3$ and formation of 0.500 M of the POCl.

Now write the equilibrium expression

$$K_c = \frac{[POCl][Cl_2]}{[POCl_3]}$$

Substitute the equilibrium concentrations from the data table

$$0.248 = \frac{(X+0.500)(0.500)}{(2.50)}$$

Solving

$$0.620 = 0.500X + 0.25$$

$$X = 0.740 \text{ M POCl}$$

The problem is not completed, however, since this is the equilibrium concentration of POCl and you were asked to find the amount added.

moles of POCl added = 0.740 M x 0.500 liters

<u>moles of POCl added =0.370 moles</u>

10. First, as usual, write the data table, but this time it's based on pressures.

	$P(I_2)$	$P(I)$
Observed pressure	0.21 atm	0.030 atm

Insert these values into the reaction quotient

$$Q = \frac{P_I^2}{P_{I_2}}$$

$$Q = \frac{(0.030 \text{ atm})^2}{(0.21 \text{ atm})}$$

$$Q = 4.3 \times 10^{-3}$$

a. Since $Q > K_p$, <u>the system is not in equilibrium.</u>
As equilibrium is approached, since $Q > K$, the equilibrium will shift to the left, and the pressure of I_2 will increase.

b. <u>The system may attain equilibrium more rapidly, but the position of the equilibrium will not change.</u>

11. Based on the slow step, the rate expression is

$$\text{rate} = k_2[I]^2[H_2]$$

To compare this rate expression with the one given, it must be converted so that it's in terms of the original reactants. The proposed mechanism indicates that I is formed rapidly from I_2, and the rate of that reaction is

$$\text{rate of production of I} = k_1[I_2]$$

and since I is removed very slowly, most of it will be converted back to I_2 before it can react with the H_2. This rate is

$$\text{rate of reversion of I to } I_2 = k_{-1}[I]^2$$

At equilibrium, the forward and reverse rates must be equal, and so

$$k_1[I_2] = k_{-1}[I]^2$$

Rearranging this expression

$$[I]^2 = \frac{k_1[I_2]}{k_{-1}}$$

Now substitute this value for $[I]^2$ into the rate expression obtained for the slow step,

$$\text{rate} = k_2[I]^2[H_2] = k_2[H_2] \times \frac{k_1[I_2]}{k_{-1}}$$

Combining all of the constants to produce one value

$$rate = k[H_2][I_2]$$

Since this is the experimentally obtained rate law, this mechanism is reasonable.

PRACTICE TEST (55 min.)

1. Consider the following two equilibrium reactions

Reaction 1 $2\ HI(g) \rightleftharpoons H_2(g) + I_2(g)$

Reaction 2 $1/2\ H_2(g) + 1/2\ I_2(g) \rightleftharpoons HI(g)$

If the equilibrium constant for reaction 1 is K_1, and the equilibrium constant for reaction 2 is K_2, what is the mathematical relationship between K_1 and K_2?

a. $K_1 = K_2^2$ b. $K_1 = 1/K_2$

c. $K_1 = -K_2$ d. $K_1 = 1/K_2^2$

2. At 275°C, $K_p = 1.14 \times 10^3\ atm^2$ for the equilibrium system

$$CH_3OH(g) \rightleftharpoons CO(g) + 2\ H_2(g)$$

What is the value of K_c for this reaction under these conditions?

3. At 400°C, $K_c = 7.0$ for the equilibrium

$$Br_2(g) + Cl_2(g) \rightleftharpoons 2\ BrCl(g)$$

If 0.060 moles of bromine gas and 0.060 moles of chlorine gas are introduced into a 1.00 liter container at 400 degrees Celsius, what is the concentration of BrCl when sufficient time has passed so that equilibrium is established?

4. The following mixture is at equilibrium in a closed container

$$NH_4HS(s) \quad \rightleftharpoons \quad NH_3(g) + H_2S(g) + \text{energy}$$

Listed on the left below are several changes that may affect the equilibrium composition of this system. For each separate change indicated, place an X in the appropriate space on the right to indicate whether that single change will increase, decrease, or have no effect on the concentration of hydrogen sulfide gas present in the system after equilibrium has been restored.

	increase [H_2S]	decrease [H_2S]	no effect
a. Add NH_3 to the system	_____	_____	_____
b. Add NH_4HS to the system	_____	_____	_____
c. Increase the temperature	_____	_____	_____
d. Add an inert gas to increase the pressure	_____	_____	_____
e. Add H_2S to the system	_____	_____	_____

5. At $2727°C$, $K_c = 0.37$ for the reaction

$$Cl_2(g) \quad \rightleftharpoons \quad 2 \, Cl(g)$$

At equilibrium in a closed container, the pressure of $Cl_2 = 0.86$ atmospheres. What is the pressure of $Cl(g)$ in this container?

6. Although coal was initially the major source of raw materials for the chemical industry, it became

less important as petroleum and natural gas became readily available. As the oil supply decreases, coal is expected to again become important. The gasification of coal will probably play an important role, and a reaction called the water-gas shift reaction is an important step in our current efforts to convert coal into gaseous products.

The water-gas shift reaction is shown below, and in one version of this process the equilibrium constant for this reaction has a value of 2.23.

$$CO(g) \quad + \quad H_2O(g) \quad \rightleftharpoons \quad CO_2(g) \quad + \quad H_2(g)$$

The table below lists the concentrations of these four reactants in a number of experiments. In each case, compare the value of the reaction quotient with the value of the equilibrium constant (2.23) to determine if the system is at equilibrium, and if it is not at equilibrium, to determine which way the reaction will probably proceed.

	Concentrations				Prediction
	[CO]	[H$_2$O]	[CO$_2$]	[H$_2$]	
a.	0.951	0.432	1.38	0.773	_____
b.	0.389	0.182	0.510	0.309	_____
c.	0.749	0.356	0.960	0.567	_____

7. If 0.800 moles of HI gas is placed in a 2.00 liter container and allowed to come to equilibrium at a constant temperature, it will dissociate according to the equation

$$2\ HI(g) \quad \rightleftharpoons \quad H_2(g) \quad + \quad I_2(g)$$

When equilibrium is attained, the concentration of iodine gas in the container is 0.100 Molar. What is the value of K_c for this reaction under these conditions?

8. Hydrogen sulfide gas is placed in a 2.00 liter

container at an elevated temperature and reacts according to the equation

$$2 \ H_2S(g) \ \rightleftharpoons \ 2 \ H_2(g) \ + \ S_2(g)$$

Calculate the value of K_c for this reaction if the resulting gas mixture consists of 1.0 moles of H_2S, 0.20 moles of H_2, and 0.80 moles of S_2.

COMMENTS ON ARMCHAIR EXERCISES

1. What effect would you expect water in the air to have on this equilibrium? Does this explain the ability of this device to predict the weather? How accurate do you think it will be?

2. Since adding heat shifts the equilibrium towards the pink color, the equilibrium must be

energy + red color $\rightleftharpoons$ pink color

This suggests that the reaction is endothermic as written.

3. You can't increase the production of the ethene, but you can make the gas concentration higher by placing the fruit in a paper bag, so that the released ethene will accumulate and hasten the ripening process. Can you explain this by using le Chatelier's principle?

CONCEPT TEST ANSWERS

1. equilibrium constant expression 2. equal rates
3. change in the number of moles of gas
4. a. pure solids   b. pure liquids   c. solvents in dilute solutions (These may be in any order.)
5. raised to a power equal to the multiplying factor
6. reciprocal   7. product
8. the equilibrium constant   9. a. K >> 1
10. the concentrations in the reaction quotient expression are not necessarily equilibrium concentrations.
11. reduce or counteract

PRACTICE TEST ANSWERS

1. **(L2)** d. $K_1 = 1/K_2^2$
2. **(L2)** $K_c = 0.565 \, M^2$
3. **(L5)** [BrCl] = 0.047 M
4. **(L6)**

	increase [H_2S]	decrease [H_2S]	no effect
a. Add NH_3 to the system		X	
b. Add NH_4HS to the system			X
c. Increase the temperature		X	
d. Add an inert gas to increase the pressure		X	
e. Add H_2S to the system	X		

5. **(L5)** $P_{Cl} = 8.9$ atm
6. **(L3)**
a. Q = 2.61, since Q > K the reaction will shift to the left.
b. Q = 2.23, since Q = K the reaction is at equilibrium.
c. Q = 2.04, since Q < K the reaction will shift to the right.
7. **(L4)** $K_c = 0.25$
8. **(L4)** $K_c = 0.0162$

CHAPTER 17
PRINCIPLES OF REACTIVITY:
THE CHEMISTRY OF ACIDS AND BASES

CHAPTER OVERVIEW

A common way to organize the study of chemical reactions is to identify a small number of reaction types, then use this as a basis for describing the reactions that fall into those categories. Three general types of inorganic reactions seem to be common in nature, acid-base, precipitation, and oxidation-reduction. The next few chapters discuss these cases in some detail.

There are several theories which can be used to define acid-base reactions, including those proposed by Arrhenius and Bronsted-Lowry. These theories are relatively simple but can be quite useful in aqueous solutions.

A particular strength of the Bronsted-Lowry theory is that it provides a model to better understand acid and base strength. According to this idea, when an acid or base is dissolved in water, the strength (or degree of ionization) will be determined by the competition between a set of conjugate acid-base pairs. This allows for a general ordering of acid and base strength in water, as seen in Tables 17.3 or 17.4.

The most common way to represent the acidity (or basicity) of a solution is in terms of the pH (or pOH). Making these determination is fairly easy for strong acids and bases, because they are highly ionized. Weak acids and bases require equilibrium calculations and so may be somewhat more difficult. To further complicate the issue, some salts are weakly acidic or basic because they include a strong conjugate ion. The essential first step in all of these problem types is recognizing what type of compound is involved.

A third acid-base theory, the Lewis theory, is more widely usable than the two theories discussed thus far. This theory is quite helpful when trying to describe the formation of complex ions in aqueous solutions as well as those cases where water is not the solvent.

LEARNING GOALS

1. Be familiar with the auto-ionization of water and the role it plays in our understanding of acids and bases. Know the Bronsted definitions of acids and bases, be able to identify the conjugate acid-base pairs in an acid-base reaction. (Secs. 17.1, 17.2, and 17.3)

2. Be able to determine the relative strength of acids or bases, including recognizing strong acids, strong bases, weak acids, and weak bases, using Table 17.3. (Sec. 17.4 and 17.5)

3. The equilibrium expression for the auto-ionization of water provides an excellent method for relating the hydrogen ion and hydroxide ion concentrations to each other.

$$K_w = [H_3O^+][OH^-]$$

This will be useful in many problems. Understand also how the relationship between acidity and basicity can also be represented using the pH scale. (Sec. 17.6)

4. Many of the problems in this chapter are based on weak acid and weak base equilibria. Degree of ionization may be indicated by the percent dissociation, ionization constant, or pH. If the initial concentration of a weak acid or base and any one of these three quantities is given, be able to calculate the missing values. This may require a review of the methods for solving quadratic equations. (Sec. 17.7)

5. Some salts may also be weakly acidic or basic when dissolved in water. This type of behavior is called hydrolysis, and the resulting pH can be calculated by methods much like those used for other weak acids and weak bases. (Sec. 17.8)

6. Be able to do calculations concerning the ionization of polyprotic acids and also understand the relationship between acid-base behavior and

molecular structure. (Sec. 17.9 and 17.10)

7. Understand the Lewis theory of acids and bases, recognize how it differs from the Bronsted theory, and use it to identify acids, bases and adducts in chemical reactions. (Sec. 17.11)

ARMCHAIR EXERCISES

1. The recommended treatment when concentrated acid is spilled on the skin is to rinse the affected area thoroughly with water. Why would concentrated base be an extremely dangerous substitute for water in this process? HINT: The base would neutralize the acid. Is this really desirable?

2. Farmers will sometimes talk about sweet and sour soil when discussing the acidity. What do you think they mean by "sour soil?"

3. In the book *Chemistry: A Systematic Approach* by Harry H. Sisler, et al, it is pointed out (pg. 586) that in a human body the pH of blood in the veins is slightly lower than the pH of the blood in the arteries. Based on what you know about the function of the blood, can you explain this observation?

4. When plant material is burned, one of the products that remains in the ashes is potassium carbonate. Would you expect potassium carbonate to be acidic or basic? Explain why you expect this to be true.

As Nathan Shalit points out in his book, *Cup and Saucer Chemistry,* you can test this easily by treating the ashes from an ash tray with water, after throwing away any cigarette ends or matches. Carefully decant off the water from the ashes, then test the solution with litmus paper or phenolphthalein.

5. Some acids and bases are normally solids, but if you try to test the solid with litmus paper there is no change in color. Only when water is added do you observe the characteristic color changes. Why is this true?

CONCEPT TEST

1. Arrhenius proposed that a substance producing _____ as one of the products of ionic dissociation in water should be called an acid, and a substance producing _____ should be called a base.

2. The Bronsted definition of acids and bases states that an acid is a(n) _____, and a base is a(n) _____.

3. Acids capable of donating more than one proton are called _____ acids; substances that can act as either a Bronsted acid or base are called _____.

4. A pair of compounds or ions that differ by the presence of one H^+ unit is called a _____.

5. Identify the conjugate acid-base pairs.

$$H_2PO_4^- \ + \ H_2O \ \rightleftharpoons \ H_3O^+ \ + \ HPO_4^{2-}$$

6. Bronsted acids that ionize 100% in water are called _____ acids; bases that ionize 100% in water are called _____ bases.

7. The Bronsted model indicates that the stronger the acid, the weaker its _____.

8. Label each compound as a strong acid, strong base, weak acid, or weak base.

a. HNO_3 _____   b. CH_3COOH _____

c. NaOH _____   d. HCl _____

9. Hydrofluoric acid, HF, is a weak acid. Based on this information, which of the following statements is most likely to be correct?

a. HF won't hurt if it's spilled on the skin.

b. F^- is a strong conjugate base.

c. HF is 100% ionized in aqueous solution.

10. _____ is the strongest acid that can exist in water. What is the strongest base that can exist in water? _____

11. At $25^{\circ}C$, pH + pOH must equal _____ for aqueous solutions.

12. A(n) _____ is a substance that changes color in some known pH range.

13. A _____ reaction is one which occurs when a salt dissolves in water and leads to changes in the concentrations of the hydronium ions and hydroxide ions.

14. For each salt, predict whether an aqueous solution of that salt will have a pH greater than,

310

less than, or equal to 7.

a. KBr   b. LiF   c. NaHSO$_3$

15. For polyprotic acids, the pH of the solution depends primarily on the hydronium ion generated in the _____.

16. In general, increasing the number of oxygen atoms in an oxyacid _____ (choose from increases or decreases) the relative acid strength.

17. The Lewis definition states that an acid is a substance that can _____ a pair of electrons from another atom to form a new bond; a Lewis base is a substance that can _____ a pair of electrons from another atom to form a new bond.

18. A substance that can behave as a Lewis acid or as a Bronsted base is said to be _____. (Contrast this with the definition of amphiprotic.)

STUDY HINTS

1. Notice that the pH scale is not limited to values between 1 and 14. Although most of the solutions commonly encountered will fall into that range, values higher than fourteen and lower than one are possible. To prove this, calculate the pH of a 1 M solution of KOH.

2. Remember that the pH of a neutral solution is 7 only when the solution temperature is 25 degrees Celsius (see practice test problem 6). In a neutral solution it is always true that hydronium ion concentration equals hydroxide ion

concentration, and so this is a better definition of a neutrality in aqueous solutions.

3. As was the case in the previous chapter, a data table is an extremely valuable way to organize the information in these equilibrium problems. If it isn't already a habit to use a data table, form that habit now.

4. For many of the problems in this chapter, the first step in the solution is to identify the type of compound involved. Strong acids and bases are 100% ionized; there is no equilibrium expression. Weak acids and bases ionize slightly, so an equilibrium equation and expression is essential. In order to determine the pH of salt solutions, it's necessary to identify the ion that undergoes hydrolysis and then write the equilibrium expression and equation for that species. Failure to determine what type of substance is involved can make this problem type difficult to identify.

5. Is it possible to make a basic solution by adding a weak acid to pure water? Of course not, but in some cases the calculations will seem to indicate that the pH of an extremely weak base is less than 7 or of an extremely weak acid is more than 7. In these cases, water becomes the species that controls the pH, and the solution is essentially neutral. Always remember to check and make sure that the answer is reasonable.

PRACTICE PROBLEMS

1. (L1) For each reaction, identify the conjugate acid-base pairs and then use Table 17.2 from the textbook to predict whether the equilibrium lies predominantly to the left or to the right.

a. $H_2PO_4^-(aq) + H_2S(aq) \rightleftharpoons H_3PO_4(aq) + HS^-(aq)$

b. $HCO_3^-(aq) + NH_4^+(aq) \rightleftharpoons NH_3(aq) + H_2CO_3(aq)$

2. (L1) a. What is the conjugate base
of HPO_4^{2-}? _____

b. What is the conjugate acid
of HPO_4^{2-}? _____

3. (L7) For plants that require acidic soil,
special fertilizers are sold that include metal ion
compounds to increase soil acidity. Examine Table
17.6 in the text and predict which of the metal
ions listed there will be most effective in
lowering the soil pH.

4. (L3) Heavy water is formed by replacing normal
hydrogen atoms with a heavier hydrogen isotope,
deuterium (symbol D), that has an atomic mass of
2.0. Chemically it is very similar to normal
water, and the equation for the auto-ionization of
heavy water, D_2O, may be written

$$2\ D_2O \rightleftharpoons D_3O^+ + OD^-$$

At $25°C$, $K_w = 1.11 \times 10^{-15}\ M^2$ for this reaction. What
is the pD of a neutral solution of heavy water at
25 degrees Celsius? Assume that pD is similar to
pH, that is, $pD = -\log [D^+]$.

5. (L3) Supply the missing values in the following
table (at $25°C$):

pH	pOH	$[H_3O^+]$	$[OH^-]$
2.45	_____	_____	_____
_____	5.42	_____	_____
_____	_____	_____	9.2×10^{-3}
_____	_____	6.5×10^{-5}	_____

6. **(L5)** Assuming equal concentrations of each substance in aqueous solution, arrange the following in order of increasing acidity: $NaC_2H_3O_2$, NH_4Cl, and $HC_2H_3O_2$.

least
(acidic) _____ < _____ < _____ most (acidic)

7. **(L3)** Calculate the pH and pOH of a 0.50 M solution of HNO_3.

8. **(L4)** The ionization equation for hydrazoic acid HN_3, is

$$HN_3 + H_2O \rightleftharpoons H_3O^+ + N_3^-$$

At $25°C$ $[H^+] = 1.38 \times 10^{-3}$ M for a 0.10 M aqueous solution of hydrazoic acid. Calculate the K_a for this weak acid under these conditions.

9. **(L4)** If $K_b = 7.4 \times 10^{-5}$ for trimethylamine, a weak base that ionizes as shown in the equation below, calculate the pH of a 0.20 M solution of this substance.

$$(CH_3)_3N(aq) + H_2O(\ell) \rightleftharpoons (CH_3)_3NH^+(aq) + OH^-(aq)$$

314

10. (L7) For each of the following compounds, predict whether it is most likely to act as a Lewis acid or a Lewis base.

a. H_2Se _____ b. AsH_3 _____

c. Ag^+ _____ d. Ti^{4+} _____

11. (L4) Calculate the pH of a 0.05 M solution of hydrofluoric acid, HF, given the following dissociation equation:

$$HF(aq) + H_2O(\ell) \rightleftharpoons H_3O^+(aq) + F^-(aq) \qquad K_a = 7.2 \times 10^{-4}$$

PRACTICE PROBLEM SOLUTIONS

1. a. $H_2PO_4^-(aq) + H_2S(aq) \rightleftharpoons H_3PO_4(aq) + HS^-(aq)$

       base₁   acid₂   acid₁   base₂

Since H_3PO_4 is a stronger proton donor than H_2S, this equilibrium will go to the left.

b. $HCO_3^-(aq) + NH_4^+(aq) \rightleftharpoons NH_3(aq) + H_2CO_3(aq)$

       base₁   acid₂   base₂   acid₁

Since H_2CO_3 is a stronger proton donor than NH_4^+, this equilibrium will also go to the left.

2. a. PO_4^{3-} is the conjugate base of HPO_4^{2-}

    b. $H_2PO_4^-$ is the conjugate acid of HPO_4^{2-}

315

3. Al^{3+} and Fe^{3+} have the largest hydrolysis constants and are commonly used to increase soil acidity for plants.

4. Write the equilibrium expression for the equilibrium reaction given.
$$K_w = [D_3O^+][OD^-]$$

In order for the solution to be neutral,
$$[D_3O^+] \text{ must equal } [OD^-]$$

and so the equilibrium expression becomes
$$K_w = [D_3O^+]^2$$

Inserting the available values

$$1.11 \times 10^{-15} = [D_3O^+]^2$$

or
$$[D_3O^+] = 3.33 \times 10^{-8}$$

Substitute this value into the pD expression given

$$pD = -\log[D_3O^+] = -\log(3.33 \times 10^{-8})$$

$$\underline{pD = 7.48}$$

5.

pH	pOH	$[H_3O^+]$	$[OH^-]$
2.45	11.55	3.5×10^{-3}	2.8×10^{-12}
8.58	**5.42**	2.6×10^{-9}	3.8×10^{-6}
11.96	2.04	1.1×10^{-12}	**9.2×10^{-3}**
4.19	9.81	**6.5×10^{-5}**	1.5×10^{-10}

6. First, identify the acidic species supplied by each compound. $HC_2H_3O_2$ is itself a weak acid, NH_4^+ and Na^+ are the acidic conjugate species in the other two compounds. Sodium ion doesn't appear on the table, but since NaOH is a strong base, its conjugate acid, sodium ion, must be a weak conjugate. From the chart, acetic acid is a stronger acid than ammonium ion. Therefore

$$NaC_2H_3O_2 < NH_4Cl < HC_2H_3O_2 \text{ (most acidic)}$$

7. Nitric acid is a strong acid and so it is 100% ionized.

$$HNO_3 + H_2O \rightarrow H_3O^+ + NO_3^-$$

Thus, the concentration of hydrogen ion is equal to the initial concentration of HNO_3.

$$[H_3O^+] = 0.50 \text{ M}$$

Solving for pH

$$pH = - \log[H_3O^+]$$

$$= - \log(0.50)$$

$$\underline{pH = 0.30}$$

There are several ways to find pOH. We will use the relationship

$$pH + pOH = 14.00$$

$$pOH = 14.00 - pH = 14.00 - 0.30$$

$$\underline{pOH = 13.70}$$

8. Since both H^+ and N_3^- are produced in equal molar amounts,

$$[N_3^-] = [H_3O^+]$$

The $[H_3O^+]$ is given, so

$$[N_3^-] = 1.38 \times 10^{-3} \text{ M}$$

The concentration of $[HN_3]$ that dissociated = $[H_3O^+]$ formed, and so

$$[HN_3] = 0.10 - 1.38 \times 10^{-3}$$

The amount subtracted fails to make a difference in the last significant figure, and so may be ignored. Now place the values we have obtained in a data table.

	$[HN_3]$	$[H_3O^+]$	$[N_3^-]$
equilibrium concentration	0.10	1.38×10^{-3}	1.38×10^{-3}

Write the equilibrium expression

$$K_a = \frac{[H_3O^+][N_3^-]}{[HN_3]}$$

$$K_a = \frac{(1.38 \times 10^{-3})(1.38 \times 10^{-3})}{(0.10)}$$

$$K_a = \underline{1.9 \times 10^{-5}}$$

9. Set up the data table for the data given

	$[(CH_3)_3N]$	$[(CH_3)_3NH^+]$	$[OH^-]$
Initial concentration	0.20	0	0
Change on going to equilibrium	-X	X	X
Equilibrium concentration	0.20 -X	X	X

Write the equilibrium expression,

$$K_b = \frac{[(CH_3)_3NH^+][OH^-]}{[(CH_3)_3N]}$$

Insert the values from the data table

$$7.4 \times 10^{-5} = \frac{X^2}{(0.20 - X)}$$

Test to determine if the X value can be ignored:

$$100 \times K_b = 7.4 \times 10^{-3}$$

Since 0.20 is greater than 100 x K_b, X can be ignored.

$$X^2 = 0.20 \times 7.4 \times 10^{-5}$$

$$X = [OH^-] = 3.8 \times 10^{-3} \text{ M}$$

$$pOH = -\log[OH^-] = -\log(3.8 \times 10^{-3})$$

$$pOH = 2.41$$

$$pH = 14.00 - pOH = 14.00 - 2.41$$

$$\underline{pH = 11.59}$$

10. a. H_2Se Lewis base   b. AsH_3 Lewis base

    c. Ag^+ Lewis acid   d. Ti^{4+} Lewis acid

11. Set up the data table as in the previous problem.

	[HF]	$[H_3O^+]$	$[F^-]$
Initial concentration	0.050	0	0
Change on going to equilibrium	-X	+X	+X
Equilibrium concentration	0.050 - X	X	X

Next, write the equilibrium expression and insert the concentrations from the data table.

$$K_a = \frac{[H_3O^+][F^-]}{[HF]}$$

$$7.2 \times 10^{-4} = \frac{X^2}{(0.050 - X)}$$

Notice that since 0.050 is not greater than 100 x K_a, X cannot be ignored! We must use the quadratic equation, which has the general form

$$aX^2 + bX - c = 0$$

After multiplying and rearranging our expression we obtain

$$X^2 + (7.2 \times 10^{-4})X - (3.6 \times 10^{-5}) = 0$$

By comparison with the standard quadratic form

$$a = 1; \quad b = 7.2 \times 10^{-4}, \text{ and } c = -3.6 \times 10^{-5}$$

Substitute these values into the equation

$$X = \frac{-b + \sqrt{b^2 - 4ac}}{2a}$$

$$X = \frac{-(7.2\times10^{-4}) + \sqrt{(7.2\times10^{-4})^2 - 4(-3.6\times10^{-5})}}{2\cdot1}$$

$$X = [H_3O^+] = 5.65\times10^{-3} \text{ M}$$

If X had been ignored instead of using the quadratic, the answer obtained would have been $[H_3O^+] = 6.0\times10^{-3}$, so the difference is significant.

Solving for the pH, using $pH = -\log[H_3O^+]$

$$pH = -\log(5.65\times10^{-3})$$

$$\underline{pH = 2.25}$$

PRACTICE TEST (45 min.)

1. For each reaction, identify the conjugate acid-base pairs and then use Table 17.3 from the textbook to predict whether the equilibrium lies predominantly to the left or to the right.

a. $H_2CO_3(aq) + CN^-(aq) \rightleftharpoons HCO_3^-(aq) + HCN(aq)$

b. $H_2O(aq) + NH_3(aq) \rightleftharpoons NH_2^-(aq) + H_3O^+(aq)$

2. a. The conjugate acid of HSO_4^- is _____.

b. The conjugate base of HSO_4^- is _____.

3. The pOH of a 0.25 M aqueous solution of an unknown weak base, which we will designate BOH, is 3.10 at 25 degrees Celsius. What is the value of K_b for the equilibrium

$$BOH \rightleftharpoons B^+ + OH^-$$

4. Using Table 17.3 in the textbook, arrange these compounds in order of increasing strength of the conjugate base: HCl, $HC_2H_3O_2$, and HCN.

5. Calculate the pH of a 0.10 M solution of $Ba(OH)_2$

6. The value of K_w for water at $50^\circ C$ is 5.48×10^{-14} M^2. What is the pH of a neutral aqueous solution at 50 degrees Celsius?

7. What is the pH of a 0.30 Molar solution of the salt NaF, if the hydrolysis of fluoride ion in water occurs according to the equation:

$$F^- \;+\; H_2O \;\rightleftharpoons\; HF \;+\; OH^- \qquad K_b = 1.4 \times 10^{-11}$$

8. At $25^\circ C$, a 0.200 M solution of a certain unknown weak acid, HA, is 1.5% ionized. What is the K_a value for the equilibrium

$$HA \;+\; H_2O \;\rightleftharpoons\; H_3O^+ \;+\; A^-$$

9. Given that $K_a = 6.3 \times 10^{-5}$ for the reaction

$$C_6H_5COOH + H_2O \;\rightleftharpoons\; C_6H_5COO^- \;+\; H_3O^+$$

Calculate how many grams of benzoic acid must be dissolved in 150 milliliters of water in order to produce a solution having a pH of 2.50.

COMMENTS ON ARMCHAIR EXERCISES

1. Since the reaction of a strong acid and a strong base is highly exothermic, trying to use a strong base to neutralize an acid spill would release large amounts of heat, which would cause even more damage than that already caused by the acid.

2. What taste is associated with acids? Does this give you a clue about the nature of sour soil?

3. The blood not only transports oxygen throughout the body; it also carries away products of metabolism, like carbon dioxide. The blood in the veins is particularly rich in carbon dioxide. Based on your knowledge of the chemistry of aquesous solutions of carbon dioxide, does this provide an explanation for the observation?

CONCEPT TEST ANSWERS

1. hydrogen ion, hydroxide ion
2. proton donor, proton acceptor
3. polyprotic, amphiprotic
4. conjugate acid-base pair
5. $H_2PO_4^-$ and HPO_4^{2-} are one pair, H_2O and H_3O^+ are the other pair.
6. strong, strong
7. conjugate base
8. HNO_3 and HCl are strong acids; HOAc is a weak acid, and NaOH is a strong base.
9. If HF is a weak acid, F^- must be a strong conjugate base. Actually, HF is very corrosive, and by definition weak acids must be only about 1% ionized.
10. hydronium ion, hydroxide ion
11. 14.00
12. indicator
13. hydrolysis
14. a. equal to 7, b. > 7, c. < 7
15. first ionization step
16. increases
17. accept (acid), donate (base)
18. amphoteric

PRACTICE TEST ANSWERS

1. (L1)a. $H_2CO_3(aq) + CN^-(aq) \rightarrow HCO_3^-(aq) + HCN (aq)$

            $acid_1$ $base_2$ $base_1$ $acid_2$

Since H_2CO_3 is a stronger acid than HCN, the equilibrium goes to the right.

b. $H_2O(aq) + NH_3(aq) \rightarrow NH_2^-(aq) + H_3O^+(aq)$

    $base_1$ $acid_2$ $base_2$ $acid_1$

Since H_3O^+ is a stronger acid than NH_3, the equilibrium is shifted to the left.

2. (L1) a. H_2SO_4 b. SO_4^{2-}
3. (L4) 2.5×10^{-6}

4. **(L2)** (weakest) $Cl^- < C_2H_3O_2^- < CN^-$ (strongest)
5. **(L3)** pH = 13.3
6. **(L3)** pH = 6.63
7. **(L4)** pH = 8.31
8. **(L4)** $K_a = 4.6 \times 10^{-5}$
9. **(L4)** 2.9 grams

CHAPTER 18
PRINCIPLES OF CHEMICAL REACTIVITY:
REACTIONS BETWEEN ACIDS AND BASES

CHAPTER OVERVIEW

Titrations have long been an important method of chemical analysis, and even though instrumental methods have replaced some of these determinations in the industrial laboratory, many natural processes are best understood as being the result of naturally occurring titrations. These natural titrations range from the current composition of the oceans and the reason why you have an upset stomach when you eat spicy food. Thus, the chemistry of titrations is very important.

You have previously learned to classify acids and bases as being either strong or weak. These categories prove extremely useful for understanding many titration problems. In fact, when preparing to work a titration problem, the first step you should take is to identify the type of acid and base involved. Once this identification is made, it's usually rather easy to make a rough sketch of the pH vs. volume curve for the titration. Calculating the pH at some point in the titration may be more difficult but is still based on the original classification of the reactants.

This chapter also introduces a special application of the common ion effect, a buffer. Buffers are commonly used in the laboratory and in many natural systems to control pH. Remember that a buffer is just a new name for a familiar problem.

LEARNING GOALS

1. Be able to calculate the pH at the equivalence point in a titration involving a strong acid and a weak base, a weak acid and a strong base, or a strong acid and a strong base. (Sec. 18.1)

2. IN order to follow the pH changes during a titration that involves a weak acid or base, it is

essential to understand the common ion effect. In addition to simple acid-base titrations, common ion or buffer solutions, as they are commonly called, also have many other practical applications. (Secs. 18.2 and 18.3)

3. Be able to draw a simple graph showing the relationship between pH and volume of titrant in an acid-base titration and also be able to select a proper indicator based on the titration graph. (Sec. 18.4 & 18.5)

ARMCHAIR EXERCISES

1. As you have learned, acetic acid is a weak acid and ammonia is a weak base. Neither of these substances ionizes very much, and so they are both weak electrolytes. If you mix equal amounts of acetic acid and ammonia, however, the resulting solution is neutral, but the conductivity is much higher. Why does the degree of ionization increase so much when you combine two materials that are both weak electrolytes?

2. Suppose that you have 500 mL of each of the following 1 Molar solutions: acetic acid, ammonia, hydrochloric acid, and sodium hydroxide. In addition, you have a 500 mL beaker and a 100 mL graduate cylinder. Can you use only these supplies to prepare a buffer having a pH of approximately 5?

3. A common remedy for headaches and upset stomach consists of a mixture of aspirin, sodium hydrogen carbonate, and citric acid (which is a weak acid). When dropped into water, this tablet produces bubbles of carbon dioxide (Can you guess what this product is?). The aspirin is obviously helpful for a headache, but in what way does this product provide relief from an upset stomach?

CONCEPT TEST

1. Acid-base reactions always proceed in the

direction of the _____ (choose from

stronger or weaker) acid-base pair.

2. The net ionic equation for the reaction of a strong acid and a strong base will be the union of the _____ ion and the _____ ion to give water.

3. Will the pH be greater than 7, less than 7, or equal to 7 at the equivalence point when a strong acid is titrated with a weak base? _____

4. When an acid and a base react in aqueous solution, if one of the reactants is weak and the other is strong, the pH of the solution after mixing equal volumes of acid and base is controlled by the _____ of the weak acid or base.

5. Will the pH be greater than 7, less than 7, or equal to 7 at the equivalence point when sodium hydroxide is titrated with acetic acid? _____

6. Will an acidic or a basic solution result if equal molar amounts of a weak acid and a weak base are mixed, and the K_b of the base is larger than the K_a of the acid? _____

7. If sodium acetate, a salt, is added to a solution of acetic acid, a weak acid, will the pH of the resulting solution be higher, lower, or the same as that of the original acetic acid solution?

8. The combination of a weak acid and its conjugate base or a weak base and its conjugate acid is used to prepare a special kind of solution called a(n) _____, because it will resist the change of pH when a strong acid or base is added.

9. When the presence of the conjugate partner affects the degree of ionization of a weak acid or base, this is called the _____.

10. The defining equation for pK_a is

$$pK_a = \text{_____}.$$

11. The Henderson-Hasselbalch equation shows clearly that the pH of the solution of a weak acid and its conjugate base is controlled primarily by the value of _____ for the weak acid.

12. What effect does simple dilution (i.e. addition of more solvent) have on the pH of a buffer? _____

13. An indicator changes color when the hydrogen ion concentration is equal to the _____ of the indicator.

14. When titrating a weak base with a strong acid, choose an indicator that changes at a pH of approximately _____.

15. In an acidic buffer solution,

$$[H_3O^+] = \underline{\hspace{6cm}}.$$

STUDY HINTS

1. One of the reasons that students find acid-base titration problems to be especially difficult is the fact that most problems really consist of three separate components. It is probably easiest to think of the process as involving three steps that must usually be accomplished in order to solve the problem.

a. Determine the initial moles of acid and moles of base and then determine what acidic or basic species are present and how many moles of each is present. Be sure to identify each remaining species as a strong acid, strong base, weak acid, weak base, or salt. This information is critical in part c.

b. Once the moles of each species present is known, it is necessary to calculate the concentration of each species. The final volume of the solution is the sum of the two (or more) components in the titration, and so be sure to use this new volume to determine the concentrations.

c. Based on the nature of the species remaining, determine whether the problem is a strong acid ionization, weak acid ionization, strong base ionization, weak base ionization, buffer, or hydrolysis problem. This should clearly identify the calculations necessary in the final step.

Breaking titration problems down into this three step process may make it easier to be successful with this type of problem.

2. There are some special cases that can greatly simplify the calculations in acid-base titrations. Learning to recognize these situations will sometimes save a great deal of work.

a. The pH is always 7 at the equivalence point

for the titration of a strong acid and a strong base.

b. If a titration is at the end point, and if both reactants have the same concentration, the concentration of the products will be half as great as that of the original reactants.

Watch for examples of these situations in the problems in this chapter.

3. It should be obvious that the first step in solving a titration or buffer problem is to write the equation for the equilibrium. It is crucial that the correct equation be selected. In most cases it will simply consist of a weak acid or base (or a conjugate acid or base) ionizing in water. Don't try to make the equations too complicated. This is a case where the simplest approach is almost always the best.

4. Suppose it may be necessary to set up a problem involving the addition of a strong acid to a buffer consisting of a weak acid, HA, and its salt, NaA. The equilibrium for the buffer would be

$$HA + H_2O \rightleftarrows H_3O^+ + A^-$$

Even though it might appear reasonable to add the concentration of the strong acid to the hydrogen ion that is already present, it is usually easier to work the problem by first allowing the acid to react with the conjugate base, A^-.

5. In the previous chapter, a rule of thumb was defined to determine whether X could be neglected when it was added to or subtracted from a much larger number. This rule used the equilibrium constant as a rough indication of the size of X. What would happen if the unknown, X, were an amount of reagent added to the solution separately rather than being the result of the equilibrium? Under these circumstances, it's no longer possible to use this rule of thumb as a justification for neglecting X. This type of problem is sometimes called a "Big X" problem, to emphasize that X cannot be neglected in the normal way. Be sure to

carefully examine the examples of this type of problem that are included in both the practice problems and the practice test.

PRACTICE PROBLEMS

1. (L3) a. Draw a rough sketch of the titration graph that will result if KOH is added to HNO_3. b. Draw the titration graph that will result when KOH is added to formic acid, HCOOH.

2. (L1) Calculate the pH that would result if 25.0 mL of 0.200 M HNO_3 is added to 20.0 mL of 0.250 M KOH.

3. (L1) In a certain titration, 20.0 mL of 0.100 M hydrochloric acid is added to 50.0 mL of 0.150 M ammonia. What is the pH of the resulting solution? (K_b = 1.8x10^{-5} for NH_3)

330

4. (L2) What is the pH of a solution prepared by dissolving 13.4 grams of CH_3NH_3Cl in 250.0 mL of 0.210 M CH_3NH_2, methylamine. Assume there is no change in solution volume when the salt of methylamine is added. ($K_b = 5.0 \times 10^{-4}$ for CH_3NH_2)

5. (L2) How many grams of $NaC_2H_3O_2$ must be added to 1.00 liter of 0.200 M acetic acid in order to prepare a buffer having a pH of 5.00. Assume that there is no change in volume when the solid sodium acetate is added. ($K_a = 1.8 \times 10^{-5}$ for acetic acid)

6. (L2) A buffer solution has been prepared by dissolving 0.500 moles of sodium formate, NaHCOO, in 1.00 liter of 0.500 Molar formic acid, HCOOH.
a. What is the pH of this solution? ($K_a = 1.8 \times 10^{-4}$ for formic acid.)

b. Calculate the pH that will result if 0.010 mole of KOH is added to 1.00 L of the buffer solution described in part a.

7. **(L2)** The initial pH is 9.08 for a solution formed by mixing 0.300 moles of solid NH_4Cl into a 0.200 M solution of ammonia having a total volume of 1.00 liter. How many moles of HCl must be added to this solution to decrease the pH by 1.00 unit. ($K_b = 1.8 \times 10^{-5}$ for ammonia.)

8. Calculate the pH at the equivalence point of a titration in which 50.0 mL of 0.300 M HF is added to 50.0 mL of 0.300 M KOH. ($K_b = 1.4 \times 10^{-11}$ for F^-.)

PRACTICE PROBLEM SOLUTIONS

1.

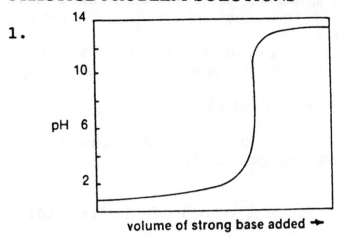

Figure 18.1 Titration curve for KOH and HNO_3

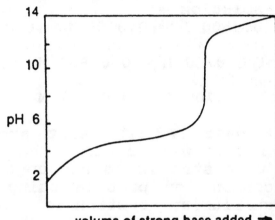

Figure 18.2 Titration curve for the addition of KOH to HCOOH.

2. First, calculate the moles of acid and base.

mol of HNO_3 = 0.0250 liters x 0.200 M = 0.00500 mol
mol of KOH = 0.0200 liters x 0.250 M = 0.00500 mol

Moles of acid equals moles of base, and so the titration is at the equivalence point. Both HNO_3 and KOH are strong, and so there is no hydrolysis.

Since the acid and base are completely reacted, and there is no hydrolysis, the solution is neutral and pH = 7.

    The pH at the end point is 7.00

3. Calculate how many moles of each substance have been used.

Mol of HCl = 0.0200 L x 0.100 M = 0.00200 mol HCl
Mol of NH_3 = 0.0500 L x 0.150 M = 0.00750 mol NH_3

Moles of NH_3 remaining = 0.00550 moles

In addition, the reaction of 0.00200 moles of HCl and 0.00200 moles of NH_3 produces 0.00200 moles of NH_4Cl.

Next, determine the concentration of the remaining species.

The total volume of solution =
0.0200 L + 0.0500 L = 0.0700 L

$[NH_3]$ = 0.00550 mol/0.0700 L = 0.07857 M

$[NH_4Cl]$ = 0.00200 mol/0.0700 L = 0.02857 M

Notice that a weak base and its salt are the remaining substances that will control the pH of the solution. The next step is to set up a data table and work a common ion problem using this information. The equilibrium reaction is

$$NH_3(aq) + H_2O(\ell) \rightleftharpoons NH_4^+(aq) + OH^-(aq)$$

Next, prepare a data table.

	$[NH_3]$	$[NH_4^+]$	$[OH^-]$
Before ionization (M)	0.07857	0.02857	0
Change	-X	+X	+X
At equilibrium	0.07857-X	0.02857+X	X

The equilibrium expression is

$$K_b = \frac{[NH_4^+][OH^-]}{[NH_3]}$$

Substitute the values from the data table.

$$1.8 \times 10^{-5} = \frac{(0.02857+X)X}{(0.07857-X)}$$

According to the rule of thumb, X may be neglected

$$X = 4.95 \times 10^{-5} = [OH^-]$$

$$pOH = -\log [OH^-] = -\log (4.95 \times 10^{-5})$$

$$pOH = 4.31$$

$$pH = 14.00 - pOH = 14.00 - 4.31 =$$

$$pH = \underline{9.69}$$

4. Since the components are a weak base and a salt of the weak base, this is a common ion problem.

First, calculate the concentration of the CH_3NH_3Cl.

Moles CH_3NH_3Cl = 13.4 g/67.45 g/mole = 0.1987 mol

Molarity CH_3NH_3Cl = 0.1987 mol/0.250 L = 0.7947 M

Based on the following equilibrium reaction

$$CH_3NH_2(aq) + H_2O(\ell) \rightleftharpoons CH_3NH_3^+(aq) + OH^-(aq)$$

Next, write the data table.

	$[CH_3NH_2]$	$[CH_3NH_3^+]$	$[OH^-]$
Before dissociation (M)	0.210	0.7947	0
Change	-X	+X	+X
Equilibrium concentrations	0.210-X	0.7947+X	X

Substituting into the equilibrium expression

$$K_b = \frac{[CH_3NH_3^+][OH^-]}{[CH_3NH_2]}$$

$$5.0 \times 10^{-4} = \frac{(0.7947 + X)X}{(0.210 - X)}$$

According to the rule of thumb, X may be neglected.

$$X = 1.32 \times 10^{-4} \text{ M} = [OH^-]$$

$$pOH = -\log[OH^-] = -\log(1.32 \times 10^{-4})$$

$$pOH = 3.88$$

$$pH = 14.00 - pOH = 14.00 - 3.88$$

$$\underline{pH = 10.12}$$

5. This solution consists of a weak acid and its salt, so it is a buffer problem. Begin by writing the balanced equation for the reaction.

$$HC_2C_3O_2(aq) + H_2O \rightleftharpoons H_3O^+(aq) + C_2H_3O_2^-(aq)$$

The unknown in this problem is the mass of sodium acetate added. This value can be designated to be X, but since the relationship between the magnitude of X and the equilibrium constant is not simple, it will not be possible to use the rule of thumb and decide to neglect this X. This is the type of "Big X" problem discussed in the study hints.

Given that the pH will be 5.00, therefore

$$[H_3O^+] = 1.0 \times 10^{-5}$$

Now develop a data table using this information.

	$[HC_2H_3O_2]$	$[H_3O^+]$	$[C_2H_3O_2^-]$
Initial conc.	0.200	0	X
Change when proceeding to equilibrium	-1.0×10^{-5}	$+1.0 \times 10^{-5}$	1.0×10^{-5}
Equilibrium conc.	$(0.200 - 1.0 \times 10^{-5})$	1.0×10^{-5}	$(X + 1.0 \times 10^{-5})$

Write the equilibrium expression and substitute the equilibrium concentrations from the data table.

$$Ka = \frac{[C_2H_3O_2^-][H_3O^+]}{[HC_2H_3O_2]}$$

$$1.8 \times 10^{-5} = \frac{(X + 1.0 \times 10^{-5})(1.0 \times 10^{-5})}{(0.200 - 1.0 \times 10^{-5})}$$

The concentration change when proceeding to equilibrium is obviously small when compared with the initial concentration of $HC_2H_3O_2$ and is probably small when compared with the concentration of $NaC_2H_3O_2$ added. For the time being, this value can be neglected, but this assumption must be checked when the problem is completed.

Substituting into the equilibrium expression

$$1.8 \times 10^{-5} = \frac{X(1.0 \times 10^{-5})}{(0.200)}$$

$$\underline{X = 0.36 \text{ M} = \text{concentration of } NaC_2H_3O_2 \text{ added}}$$

Notice that 1.0×10^{-5} is negligible compared with X.

To calculate how many grams of sodium acetate are needed, determine the number of moles and multiply this by the molar mass of sodium acetate.

grams of $NaC_2H_3O_2$ = 0.36 M x 1.00 L x 82.0 g/mol

$$\underline{\text{grams of } NaC_2H_3O_2 = 29.5 \text{ grams}}$$

6. a. Begin by writing the balanced equation.

$$HCOOH(aq) + H_2O(\ell) \rightleftharpoons H_3O^+(aq) + HCOO^-(aq)$$

and then set up the table of concentrations.

| | [HCOOH] | [H₃O⁺] | [HCOO⁻] |

	[HCOOH]	[H_3O^+]	[HCOO^-]
Initial conc. (M)	0.500	0	0.500
Change on proceeding to equilibrium	-X	+X	+X
Equilibrium conc. (M)	0.500-X	X	0.500+X

Substitute these values into the equilibrium expression.

$$Ka = 1.8 \times 10^{-4} = \frac{[H_3O^+][HCOO^-]}{[HCOOH]} = \frac{X(0.500+X)}{(0.500-X)}$$

Solving for hydrogen ion concentration with the assumption that X may be neglected

$$X = [H_3O^+] = 1.8 \times 10^{-4}$$

$$pH = -\log[H_3O^+] = -\log(1.8 \times 10^{-4})$$

$$pH = 3.74$$

b. The equilibrium remains the same, but in this case 0.010 moles of KOH has been added. Write the data table again, but this time allow the KOH to react with the HCOOH, decreasing the HCOOH concentration and producing more HCOO⁻ ion.

The appropriate data table is

	[HCOOH]	[H_3O^+]	[HCOO^-]
Initial conc. (M)	0.500	0	0.500
Change on adding KOH	-0.010		+0.010
Concentration after adding KOH	0.490		0.510
Change on proceeding to equilibrium	-X	+X	+X
Equilibrium conc. (M)	0.490-X	X	0.510+X

Substituting into the equilibrium expression

$$K_a = 1.8\times10^{-4} = \frac{[H_3O^+][HCOO^-]}{[HCOOH]} = \frac{X(0.510+X)}{(0.490-X)}$$

Neglecting X and solving

$$X = [H_3O^+] = 1.73\times10^{-4}$$

$$pH = -\log[H_3O^+] = -\log(1.73\times10^{-4})$$

$$\underline{pH = 3.76}$$

As expected, the buffer solution shows a very small change in pH despite the addition of a strong base.

7. This is a buffer solution, with the equation

$$NH_3(aq) + H_2O(\ell) \rightleftharpoons NH_4^+(aq) + OH^-(aq)$$

When HCl, a strong acid, is added, it will react with the conjugate base, NH_3, and the data table will show this change. The amount of HCl added is an unknown, X, but since it is not directly related to the equilibrium constant, it's not possible to neglect X. This is another "Big X" problem.

Given that the final pH is 8.08

$$[OH^-] = \text{antilog}(-pOH) = \text{antilog}(-5.92)$$

$$[OH^-] = 1.202\times10^{-6} \text{ M}.$$

Now write the data table.

	$[NH_3]$	$[NH_4^+]$	$[OH^-]$
Initial conc.	0.200	0.300	0
Change due to HCl added	-X	+X	0
Conc. after HCl added	0.200-X	0.300+X	0

Change when proceeding
  to equilibrium -1.202×10^{-6} $+1.202 \times 10^{-6}$ $+1.202 \times 10^{-6}$
Equilibrium
  conc.   0.200-X-   0.300+X+ 1.202×10^{-6}
                        1.202×10^{-6} 1.202×10^{-6}

It appears that 1.202×10^{-6}, the change upon proceeding to equilibrium is small enough so that its effect on the concentration of NH_3 and NH_4^+ can be ignored, but this assumption will need to be checked when the problem is completed.

Now write the equilibrium expression and solve using the values from the data table.

$$K_b = \frac{[NH_4^+][OH^-]}{[NH_3]}$$

$$1.8 \times 10^{-5} = \frac{(0.300+X)(1.202 \times 10^{-6})}{(0.200-X)}$$

   X = 0.169 M = concentration of HCl added

(Notice that the decision to neglect 1.202×10^{-6} was justified.)

Since the solution has a volume of 1.00 L

        Moles of HCl = 0.169 M x 1.00 L

        Moles of HCl = 0.169 moles

8. Determine the moles of each reactant present.

Moles of HF = 0.0500 L x 0.300 M = 0.0150 moles
Moles of KOH = 0.0500 L x 0.300 M = 0.0150 moles

The titration is at the equivalence point, and acidity of the solution will be determined by the hydrolysis of the salt, NaF.

Conc. of NaF = 0.0150 moles/0.100 L = 0.150 M

The F^- ion is the strong conjugate, and the equation for its hydrolysis is

$$F^-(aq) + H_2O(\ell) \rightleftharpoons HF(aq) + OH^-(aq)$$

Now set up the data table for this situation.

	$[F^-]$	$[HF]$	$[OH^-]$
Before hydrolysis (M)	0.150	0	0
Change on proceeding to equilibrium	-X	+X	+X
Equilibrium conc. (M)	(0.150-X)	X	X

Inserting these values into the equilibrium expression

$$K_b = \frac{[HF][OH^-]}{[F^-]} = \frac{x^2}{(0.150-X)} = 1.4 \times 10^{-11}$$

The rule of thumb indicates that X may be neglected.

$$X = [OH^-] = 1.45 \times 10^{-6}$$

$$pOH = -\log[OH^-] = -\log(1.45 \times 10^{-6}) = 5.84$$

$$pH = 14.00 - pOH = 14.00 - 5.84$$

$$\underline{pH = 8.16}$$

PRACTICE TEST (60 min.)

1. Calculate the pH that results when 0.25 moles of sodium acetate is dissolved in 500 mL of 0.15 M acetic acid. ($K_a = 1.8 \times 10^{-5}$ for acetic acid. Assume no volume change upon addition of solid.)

2. Calculate the pH that would result if 25.0 mL of 0.150 M formic acid, HCOOH, is added to 25.0 mL of 0.200 M NaOH. ($K_a = 1.8 \times 10^{-4}$ for formic acid.)

3. Calculate the pH at the equivalence point of a titration involving 0.30 M NH_3 and 0.30 M HCl. (K_a = 5.6×10^{-10} for ammonium ion. Notice that the pH will not depend on the volumes of acid and base.)

4. How many moles of solid NH_4Cl must be added to 1.00 liter of 0.200 NH_3 in order to produce a final solution having a pH of 9.40? (Assume that the addition of the NH_4Cl doesn't change the volume of the solution. K_b = 1.8×10^{-5} for ammonia.)

5. Calculate the pH that results when 50.0 mL of 0.200 M KOH is added to 100.0 mL of 0.200 M HF. (K_a = 7.2×10^{-4} for hydrofluoric acid.)

6. Solid sodium acetate is added to 500.0 mL of 0.300 M $HC_2H_3O_2$ until the pH of the resulting solution is 5.70 at $25°C$. Assuming that there is no volume change when the solid is added, how many grams of sodium acetate were added? (K_a = 1.8×10^{-5} for acetic acid.)

7. Which of the following pairs of compounds would be most appropriate for the preparation of a buffer having a pH of 5.0? What concentration of each compound might be used to produce exactly this pH.

a. C_5H_5N/C_5H_5NHCl K_b = 1.7×10^{-9}

b. NH_3/NH_4Cl K_b = 1.8×10^{-5}

c. $HC_2H_3O_2/NaC_2H_3O_2$ K_a = 1.8×10^{-5}

d. HF/NaF K_a = 6.8×10^{-4}

8. A balanced buffer is prepared by dissolving 1.00 mole of $NaC_2H_3O_2$ in 1.00 liter of 1.00 M $HC_2H_3O_2$ and is found to have a pH of 4.74. How many moles of solid NaOH must be added to this solution to increase the final pH to 5.75?

COMMENTS ON ARMCHAIR EXERCISES

1. When ammonia reacts with acetic acid, the product is a salt, ammonium acetate. Does what you

know about the ionization of salts enable you to explain what is happening?

2. To obtain a pH of about 5, the best choice from those available would be a buffer consisting of acetic acid and sodium acetate. You have the acetic acid but not the sodium acetate. Can you produce the sodium acetate by adding sodium hydroxide to the acetic acid? How much sodium hydroxide should you add if you wish to obtain the indicated pH?

3. The reaction of the basic sodium hydrogen carbonate and the acidic citric acid produces carbon dioxide gas. It also partially neutralizes the citric acid. Can you see that this will produce a buffer solution, which will help protect the human stomach for the excess acidity associated with overindulging?

CONCEPT TEST ANSWERS

1. weaker
3. less than 7
5. greater than 7
7. higher pH (less acidic)
9. common ion effect
11. K_a (or pK_a)
13. K_a
15. $$\frac{K_a \times [acid]}{[conjugate\ base]}$$

2. H_3O^+, OH^-
4. conjugate partner
6. basic
8. buffer
10. $-\log(K_a)$
12. none
14. eight

PRACTICE TEST ANSWERS

1. (L2) pH = 5.27
2. (L2) pH = 12.40
3. (L1) 5.04
4. (L2) 0.14 mole
5. (L1) 3.17
6. (L2) 111 grams
7. (L2) $HC_2H_3O_2/NaC_2H_3O_2$, as long as $[C_2H_3O_2^-]$ = 1.8$[HC_2H_3O_2]$, the correct pH will result.

8. (L2) 0.82 moles

CHAPTER 19
PRINCIPLES OF REACTIVITY:
PRECIPITATION REACTIONS

CHAPTER OVERVIEW

The second of the main reaction types that are normally identified in inorganic chemistry is precipitation reactions. In many cases, this category is also another example of equilibrium processes. Many of the problem types that you have learned recently are equally applicable here, such as calculating the equilibrium constant (K^{sp}) values, determining equilibrium concentrations, and predicting the effect of a common ion.

There are many different applications of this type of equilibrium, including ion separations that you may have done in the laboratory, deposition of minerals in the oceans, and even the biogeochemical cycle for elements like carbon.

LEARNING GOALS

1. Based on the equations for precipitation reactions, write the solubility product expressions for insoluble salts. (Sec. 19.1)

2. Be able to calculate the solubility product constant based on solubility information, or if given the value of the solubility product constant for a compound, be able to calculate the solubility and/or concentration of the ions produced by that compound in aqueous solution, and also be able to determine which of two salts is more soluble, based on the K_{sp} values. (Sec. 19.2 & 19.3)

3. Given the value of the solubility product for a compound and the concentrations of the component ions in solution, be able to predict whether a precipitate will form. This technique is important when using solubility differences to accomplish the separations of metal ions. (Sec. 19.4)

4. A common ion will shift the position of a

solubility equilibrium just as was observed with other equilibria. The method of calculation is similar to problem types encountered previously, so this should be relatively familiar. (Sec. 19.5)

5. Be able to evaluate the probable effectiveness of proposed methods for selective precipitation and also to develop new selective precipitation procedures based on solubilities or solubility product values. (Sec. 19.6)

6. Understand that in some cases the formation of complex ions or a change in solution pH may have a significant effect on solubility. (Sec. 19.7 & 19.8)

ARMCHAIR EXERCISES

1. An interesting example of precipitate formation is described in the book *Chemical Demonstrations: A Sourcebook for Teachers*, Volume 2, Second Edition (pg. 132) written by Lee R. Summerlin, et al. When a solution of barium hydroxide is tested with a simple conductivity meter, the solution is found to be a good conductor. If dilute sulfuric acid is added to the barium hydroxide, the conductivity steadily decreases, until the conductivity seems quite low. With further addition of the sulfuric acid, however, the conductivity increases again. Can you explain why the conductivity first decreased, then increased?

2. When performing a precipitation reaction, you will often be asked to use an excess of the precipitating agent. Does this actually help to increase the amount of precipitate formed? If so explain why this works.

CONCEPT TEST

1. The text suggests that insoluble compounds usually dissolve less than _____ mole/Liter.

2. A precipitation reaction is an exchange reaction in which one of the products is _____.

3. When the salt MgF_2 is dissolved in water, the

resulting concentration of fluoride ion will be twice as great as that of the magnesium ion. Will this relationship always be true?

_____

4. If the reaction quotient is greater than the value of K_{sp} for a given system, do you expect a precipitate to form? _____ (yes or no?)

5. When a common ion is added to a salt solution, the solubility of the salt _____. (Choose increases, decreases, or remains the same).

6. When separating ions by differences in solubility, the substance that precipitates first is the one whose _____ is first exceeded.

7. If carbonate ion is slowly added to a solution that is 0.01 M in both Ba^{2+} and Ca^{2+} ions, based on Table 19.2 in the textbook, which will precipitate first, $CaCO_3$ or $BaCO_3$? _____

8. In order to increase the solubility of $Ca(OH)_2$, is it better to use a strong acid or a strong base as the solvent? _____

9. Calcite is more soluble in the cold water of the deep oceans than it is in the warmer surface waters. Does this suggest that dissolving calcite is an exothermic or an endothermic process?

_____.

10. If the value of the reaction quotient, Q, is greater than the value of K_{sp}, the system is said to be _____.

11. Insoluble inorganic salts containing anions derived from _____ tend to be soluble in solutions of strong acids.

STUDY HINTS

1. When working common ion problems, students sometimes become confused about when to use the stoichiometric coefficients. To clarify this question, examine where the ion came from. For example, consider the addition of NaCl to a solution of $MgCl_2$. The chloride ion that results from the ionization of the magnesium chloride will be twice as great as the concentration of the dissolved magnesium chloride, but the chloride ion from the sodium chloride will be equal to the concentration of dissolved sodium chloride. Of course, if it is stated that the chloride ion concentration has a certain value, it is not necessary to multiply regardless of the source of the chloride ion.

2. For some reason, when the same stoichiometric factor is used to both multiply a concentration and raise to a power, some students are more likely to omit one or the other of these steps. Remember that the same stoichiometric factor may need to be used both in the concentrations for the data table as well as in the solubility product expression!

3. In the previous chapter, it was noted that in some cases it isn't possible to use the rule of thumb to neglect X. A similar situation is also found with solubility problems, so watch carefully for these "Big X" problems

PRACTICE PROBLEMS

1. (L1) Write the K_{sp} expression for each of the following salts:

a. AgI

b. $Cr(OH)_3$

c. Ag_3PO_4

d. Co_2S_3

2. **(L2)** The solubility of AgCl is 1.3×10^{-5} M. What is the value of K_{sp} for this compound?

3. **(L2)** Given that $K_{sp} = 6.4 \times 10^{-9}$ for MgF_2, a. calculate the solubility of MgF_2 in water in mol/L. b. calculate the solubility of MgF_2 in water in mg/L

4. **(L3)** Will a precipitate form if 0.00010 moles of solid NaCl is added to 500 milliliters of 0.0020 M $Pb(NO_3)_2$?   ($K_{sp} = 1.7 \times 10^{-5}$ for $PbCl_2$.)

5. **(L4)** Calculate the solubility (mol/L) of PbI_2 in a 0.20 Molar solution of NaI.   ($K_{sp} = 8.7 \times 10^{-9}$)

6. **(L6)** Calculate the pH required in an aqueous solution in order to reduce the solubility of $Al(OH)_3$ to 2.0×10^{-14} M. ($K_{sp} = 1.9 \times 10^{-33}$ for $Al(OH)_3$

7. **(L5)** If hydroxide ion is very slowly added to a solution containing 0.0010 M Fe^{2+} and 0.0010 M Co^{2+}, which ion will precipitate first, and what is the concentration of the first ion that remains in solution when the second ion just begins to precipitate? ($K_{sp} = 7.9 \times 10^{-15}$ for $Fe(OH)_2$ and $K_{sp} = 2.5 \times 10^{-16}$ for $Co(OH)_2$.)

PRACTICE PROBLEM SOLUTIONS

1. a. $K_{sp} = [Ag^+][I^-]$
   b. $K_{sp} = [Cr^{3+}][OH^-]^3$
   c. $K_{sp} = [Ag^+]^3[PO_4^{3-}]$
   d. $K_{sp} = [Co^{3+}]^2[S^{2-}]^3$

2. When silver chloride dissolves in water, the reaction can be represented

$$AgCl(s) \rightleftarrows Ag^+(aq) + Cl^-(aq)$$

and the K_{sp} expression for AgCl is

$$K_{sp} = [Ag^+][Cl^-]$$

Each mole of AgCl that dissolves will produce one mole of Ag^+ and one mole of Cl^-, so if 1.3×10^{-5} mol/L dissolves then

$$[Ag^+] = [Cl^-] = 1.3 \times 10^{-5}$$

Substituting into the K_{sp} expression

$$K_{sp} = (1.3 \times 10^{-5})^2$$

$$K_{sp} = \underline{1.7 \times 10^{-10}}$$

3. a. The equation for the dissolving of MgF_2 is

$$MgF_2(s) \rightleftharpoons Mg^{2+}(aq) + 2 F^-(aq)$$

and the K_{sp} expression is

$$K_{sp} = [Mg^{2+}][F^-]^2$$

If S = the solubility of this salt, then $S = [Mg^{2+}] = 2[F^-]$.

Substituting these values into the K_{sp} expression

$$K_{sp} = [Mg^{2+}][F^-]^2$$

$$K_{sp} = (S)(2S)^2 = 4S^3 = 6.4 \times 10^{-9}$$

$$\underline{S = 1.2 \times 10^{-3} M}$$

b. Now convert the solubility in mol/L to mg/L.

$$S(mg/L) = 1.2 \times 10^{-3} \text{ mol/L} \times 62.3 \text{ g/mol} \times 1000 \text{ mg/g}$$

$$\underline{S(mg/L) = 73 \text{ mg/L}}$$

4. The equation for the dissolving of the salt is

$$PbCl_2(s) \rightleftharpoons Pb^{2+} + 2Cl^-$$

And so the K_{sp} expression for $PbCl_2$ is

$$K_{sp} = [Pb^{2+}][Cl^-]^2$$

Since NaCl and $Pb(NO_3)_2$ are quite soluble, the concentration of the Cl^- and Pb^{2+} ions is equal to the concentrations of the salts that dissolved to produce these ions.

$[Pb^{2+}] = 0.0020$ M

Notice that the chloride ion concentration must be calculated from the moles of compound and volume of solution.

$[Cl^-] = 0.00010$ mol$/0.500$ L $= 0.00020$ M

[Notice that the source of the chloride ion in this case was NaCl, and so it was not necessary to double the concentration. If the chloride ion had been produced by the $PbCl_2$, the situation would have been different.]

Now substitute these values into the K_{sp} expression.

$$Q = (0.0020)(0.00020)^2 = 8.0 \times 10^{-11}$$

Since $Q < 1.7 \times 10^{-5}$, a precipitate will not form.

5. Sodium iodide is quite soluble, so assume that all of it is in solution and completely dissociated. By definition, determining the solubility of PbI_2 indicates that some solid doesn't dissolve.

Write the equations for the two compounds in water.

$$NaI(s) \rightarrow Na^+(aq) + I^-(aq)$$

$$PbI_2(s) \rightleftharpoons Pb^{2+}(aq) + 2\ I^-(aq)$$

Set up a data table for this solution, as shown on the next page, using S to represent the solubility of the lead iodide.

	$[Pb^{2+}]$	$[I^-]$
Initial concentration (M)	0	0.20
Change on proceeding to equilibrium	+S	+2S
Equilibrium concentration	S	0.20+2S

[Notice that the iodide concentration that resulted from the ionization of the NaI is not doubled, but the iodide concentration that results from the dissolving of the lead iodide is doubled. Examination of the equations for these reactions should make it obvious why this is true.]

Now write the K_{sp} expression and substitute these values.

$$K_{sp} = 8.7 \times 10^{-9} = [Pb^{2+}][Cl^-]^2 = S(0.20+2S)^2$$

The value of S is quite small, even in pure water, and the result of the common ion will be to decrease this concentration even further. It is possible to assume that 0.20+2S is approximately equal to 0.20 without causing a significant error.

The equation then becomes

$$K_{sp} = (0.20)^2 S = 8.7 \times 10^{-9}$$

$$\underline{S = 2.2 \times 10^{-7} \text{ M}}$$

Checking the assumption, it is clear that 0.20 is much larger than the value of 2S.

6. Begin by writing the equation for the dissolving of the aluminum hydroxide.

$$Al(OH)_3(s) \rightleftharpoons Al^{3+}(aq) + 3 OH^-(aq)$$

The question suggests that a base is being added to increase the pH and decrease the solubility of the $Al(OH)_3$. The hydroxide ion concentration will be a combination of the solubility of the aluminum hydroxide and this other source of base.

The best way to show this is with a data table.

	$[Al^{3+}]$	$[OH^-]$
Equilibrium conc. (M)	S	3S+X

The required solubility is 2.0×10^{-14} M, and this value can be substituted for S in the data table.

| Equilibrium conc. (M) | 2.0×10^{-14} | $X+3(2.0 \times 10^{-14})$ |

The X value cannot be neglected, since the concentration of common ion is rather large compared to the result from the dissolved compound.

If you assume that $X \gg 2.0 \times 10^{-14}$. These values can now be substituted into the K_{sp} expression.

$$K_{sp} = 1.9 \times 10^{-33} = [Al^{3+}][OH^-]^3 = (2.0 \times 10^{-14})X^3$$

$$X = 4.6 \times 10^{-7} = [OH^-]$$

$$pOH = -\log[OH^-] = -\log(4.6 \times 10^{-7}) = 6.34$$

$$\underline{pH = 7.66}$$

7. Calculate the concentration of hydroxide ion necessary to precipitate each of these ions.

For $Fe(OH)_2$,

$$K_{sp} = [Fe^{2+}][OH^-]^2$$

and so

$$[OH^-]^2 = \frac{K_{sp}}{[Fe^{2+}]} = \frac{7.9 \times 10^{-15}}{0.0010}$$

$$[OH^-] = 2.8 \times 10^{-6}$$

For $Co(OH)_2$,

$$K_{sp} = [Co^{2+}][OH^-]^2$$

or

$$[OH^-]^2 = \frac{K_{sp}}{[Co^{2+}]} = \frac{2.5 \times 10^{-16}}{0.0010}$$

$$[OH^-] = 5.0 \times 10^{-7}$$

Since the cobalt(II) requires a smaller OH⁻ concentration, it will be the first ion to precipitate.

The next question really asks what concentration of cobalt(II) will remain in solution when the iron(II) just begins to precipitate. It has already been shown that iron(II) will just begin to precipitate when $[OH^-] = 2.8 \times 10^{-6}$. How much cobalt(II) will remain in solution when the hydroxide concentration reaches that level?

Substitute the concentration of hydroxide ion when the iron just begins to precipitate, into the solubility product expression for the cobalt hydroxide to find the concentration of cobalt ion remaining at that time.

$$[Co^{2+}] \text{ as } Fe^{2+} \text{ just begins to precipitate} = \frac{K_{sp}}{[OH^-]^2}$$

$$[Co^{2+}] \text{ remaining} = \frac{2.5 \times 10^{-16}}{(2.8 \times 10^{-6})^2} = 3.2 \times 10^{-5} \text{ M}$$

Thus, when iron(II) hydroxide begins to precipitate

$$[Co^{2+}] = 3.2 \times 10^{-5} \text{ M}$$

PRACTICE TEST (45 min.)

1. Arrange the following compounds in order of decreasing solubility in water:

PbS	$K_{sp} = 8.4 \times 10^{-28}$
CdS	$K_{sp} = 3.6 \times 10^{-29}$
NiS	$K_{sp} = 3.0 \times 10^{-21}$

2. If $K_{sp} = 6.3 \times 10^{-6}$ for $PbBr_2$ in aqueous solution, calculate the solubility of this salt in water.

3. Given that $K_{sp} = 1.8 \times 10^{-10}$ for AgCl, determine whether or not a precipitate will form if 0.00010 moles of solid NaCl is added to 500 milliliters of 0.0020 M AgNO$_3$.

4. If $K_{sp} = 8.1 \times 10^{-19}$ for BiI$_3$, what is the solubility of this salt in water?

5. If $K_{sp} = 3.9 \times 10^{-11}$ for CaF$_2$, what is the Molar solubility of CaF$_2$ in a 0.20 Molar CaCl$_2$ solution?

6. Calculate the pH required in an aqueous solution in order to reduce the solubility of Mg(OH)$_2$ to 2.0×10^{-6} M. ($K_{sp} = 1.5 \times 10^{-11}$ for Mg(OH)$_2$)

7. If sulfide ion is very slowly added to a solution containing 0.0050 M Zn^{2+} and 0.0050 M Co^{2+}, which ion will precipitate first, and what is the concentration of this ion that remains in solution when the second ion just begins to precipitate? ($K_{sp} = 5.9 \times 10^{-21}$ for CoS and $K_{sp} = 1.1 \times 10^{-21}$ for ZnS.)

8. Based on the equilibrium constant values for the following reactions:

$$Cd(NH_3)_4^{2+}(aq) \rightleftharpoons Cd^{2+}(aq) + 4\ NH_3(aq) \qquad K = 1.0 \times 10^{-7}$$

$$CdS(s) \rightleftharpoons Cd^{2+}(aq) + S^{2-}(aq) \qquad\qquad K = 3.6 \times 10^{-29}$$

Calculate the value of the equilibrium constant for the reaction

$$CdS(s) + 4\ NH_3(aq) \rightleftharpoons Cd(NH_3)_4^{2+}(aq) + S^{2-}(aq)$$

COMMENTS ON ARMCHAIR EXERCISES

1. Initially, the ionization of the barium hydroxide produces enough ions to conduct the electric current. As sulfuric added, barium sulfate precipitates. When the end point of the titration is reached, the only remaining ions will be very low concentrations of barium and sulfate

ions (as indicated by the solubility product expression) and hydronium and hydroxide ions that result from the autoionization of water. This is not enough to make the solution a very good conductor. What happens as excess acid is added?

2. Once equilibrium is attained, as indicated by the presence of a precipitate, the concentrations of the ions are determined by the solubility product expression. Increasing the concentration of one of the ions included in that expression will force the other ion concentration to become smaller.

CONCEPT TEST ANSWERS

1. 0.01 Mole/Liter
2. an insoluble compound
3. Yes, unless another compound is present that can also serve as a source of magnesium or fluoride ions. For example, if the solution contains both MgF_2 and NaF, the concentration of fluoride will not be twice that of the magnesium ion.
4. yes
6. K_{sp}
8. strong acid
10. supersaturated

5. decreases
7. $CaCO_3$
9. exothermic
11. weak acids

PRACTICE TEST ANSWERS

1. (L2) NiS > PbS > CdS (least soluble)
2. (L2) 1.2×10^{-2} M
3. (L3) Yes, since $Q = 4 \times 10^{-7}$, which is greater than K_{sp}, a precipitate will form .
4. (L2) 1.3×10^{-5}
5. (L4) 7.0×10^{-6}
6. (L6) pH = 11.44
7. (L3) ZnS will precipitate at a lower sulfide ion concentration (i.e. 2.2×10^{-19}) and when CoS is about to begin to precipitate the concentration of zinc remaining in solution is 9.2×10^{-4} M
8. (L6) 3.6×10^{-22}

CHAPTER 20
PRINCIPLES OF REACTIVITY:
<u>ENTROPY AND FREE ENERGY</u>

CHAPTER OVERVIEW

Predicting whether a reaction will be product-favored or reactant-favored is not simply an interesting intellectual exercise for chemists. Using thermodynamics it is possible to predict what reactions may take place as well as the conditions under which they are most likely to occur. This can be extremely useful, since it sometimes enables a chemist in the laboratory to focus his or her attention on reactions that are most likely to occur and to ignore those reactions which are unlikely to ever produce significant products.

Much earlier in the course, you have already encountered one of the major tools chemists use to predict whether or not reactions occur, namely the enthalpy change during a process. The second important criteria is the change in entropy. Even though this seems more abstract than energy change, it is equally important for the understanding of chemical reactions.

The ultimate evaluation of the probability of a chemical reaction occurring must include both enthalpy and entropy changes, and the equation that combines these two tendencies, the Gibbs equation, is then a valuable tool in analyzing chemical reactions. Since both free energy and the equilibrium constant provide alternative ways of examining the tendency for a reaction to occur, it is reasonable to expect that there be some relationship between these two quantities, and this expectation is fulfilled.

LEARNING GOALS

1. As noted previously, energy change plays an important role in determining the spontaneity of a reaction. This chapter introduces a new state function, entropy, which is another fundamental

measure of reaction spontaneity. Be able to qualitatively predict whether entropy increases or decreases for simple processes and also be able to calculate the entropy change for changes of state and chemical reactions. (Sec. 20.1 & 20.2)

2. This chapter also introduces the third major state function that determines reaction spontaneity, Gibbs free energy. Be able to calculate free energy changes using a table of standard free energy values and the equation

$$\Delta G_f^o = \Sigma \Delta G_f^o(\text{products}) - \Sigma \Delta G_f^o (\text{reactants})$$

Also understand the relationship between the sign of the free energy change and reaction spontaneity. (Sec. 20.3)

3. Given the standard enthalpy and entropy changes for a process, know how to use the formula

$$\Delta G^o = \Delta H^o - T \Delta S^o$$

to determine the free energy change at any temperature. (Sec. 20.3)

4. Gibbs free energy is closely related to the equilibrium constant by the equation

$$\Delta G^o = - RT \ln K$$

Be able to determine the K value, given appropriate thermodynamic values, such as entropy change, enthalpy change, or free energy change. (Sec. 20.4)

ARMCHAIR EXERCISES

1. Dissolving common salt, NaCl, in water is an endothermic process. Does this observation suggest that dissolving salt in water should be product-favored or reactant-favored? Based on your experience, is this correct? If not, suggest an explanation of why the prediction based on the enthalpy change is not correct.

2. One of the odd bits of humor that is sometimes shared by chemists is the suggestion that the

purpose of thermodynamics is to put experimental chemists out of business. That is, if tables of appropriate values are available, it might be possible to predict what will happen in any chemical reaction based on thermodynamic calculations. From what you know, does it seem likely that this goal can be achieved?

3. Suppose that you place a layer of white marbles in a jar and then place a layer of blue marbles on top of the first layer. If you shake up the marbles, they will become mixed. What is the energy difference between the mixed and the unmixed configurations? What is the probability that you can return the jar to the original configuration (one separate layer of each color) by random mixing? Why is this?

CONCEPT TEST

1. A chemical reaction in which most of the reactants can eventually be converted to products, given sufficient time, is said to be _____.

2. If a reaction is reactant favored, does that mean that it does not occur at all? _____.

3. What two factors must be considered when trying to predict whether or not a particular process will be product favored?

a. _____

b. _____

4. The thermodynamic function called _____ is a measure of the matter dispersal or disorder of a system.

5. If the entropy of each element in some crystalline state is taken as _____ at 0 K, every substance has a finite entropy. This statement is called the _____.

6. Indicate whether each process below represents an increase or a decrease in entropy of the system.

a. Water evaporates from a beaker

b. Sugar dissolves in coffee.

c. Atmospheric oxygen dissolves in lake waters

d. An ice cube melts.

7. The second law of thermodynamics states that the total _____ of the universe is constantly increasing.

8. If a reaction is exothermic and also proceeds from a state of order to one of disorder, it will be _____. (Chose from product-favored or not product-favored.)

9. In order for a process to be product-favored, the sign of the free energy change must be _____ (choose from positive, negative, or zero.)

10. The standard free energy change for the formation of one mole of a compound from its elements with all reactants and products in the standard state is called the _____.

11. Elements in their standard states have free energy change values of _____.

12. In the relationship, $\Delta G^{\circ} = - RT \ln K$, _____ is the value normally used for R.

STUDY HINTS

1. Study sections 20.1 and 20.2 of the textbook carefully. They provide some valuable hints for understanding entropy better and predicting the sign of the entropy change for simple processes.

2. When working with quantities like ΔG° or ΔH°, remember that the superscript automatically indicates the temperature must be $25^{\circ}C$, even if this

value is not listed in the problem.

3. In a number of cases, it is a general rule that the value of certain thermodynamic variables is always zero for an element in its standard state under standard conditions. Remember that this is not usually true for the absolute entropy values of the elements.

PRACTICE PROBLEMS

1. (L1) By inspection, predict whether the entropy value will increase or decrease during the following processes.

a. $H_2O(\ell)$ → $H_2O(s)$

b. $O_2(g)$ → $2\ O(g)$

c. $NaCl(s)$ → $NaCl(aq)$

d. $NH_3(g)$ → $NH_3(aq)$

2. (L2) The standard free energy of formation for diamond is listed as 2.90 kJ/mole. Which is more stable, diamond or graphite? Does this agree with the common idea regarding these two substances?

3. (L1) The enthalpy of vaporization of carbon tetrachloride is 30.0 kJ/mole at it normal boiling point of 76.8°C. Calculate the entropy change that occurs when 1.0 mole of liquid carbon tetrachloride vaporizes at the boiling point.

4. (**L2**) Hydrazine, N_2H_4, and its methyl derivatives are used extensively for rocket fuels in guided missiles, space shuttles, and lunar landers. It can be used with a variety of oxidizers, including O_2, H_2O_2, and F_2. The equation for one such reaction is

$$N_2H_4(\ell) + 2\ H_2O_2(\ell) \rightarrow N_2(g) + 4\ H_2O(g)$$

Using the free energy values provided below, calculate the standard free energy change for this reaction.

Standard Free Energies of Formation (kJ/mol)

$N_2H_4(\ell)$ 149.2; $H_2O_2(\ell)$ −120.4; $H_2O(g)$ −228.6

5. (**L3**) As shown in the equation below, nitrogen oxide, NO, is a common pollutant that is produced by the reaction of nitrogen and oxygen gases. Since these gases are the major components of air, nitrogen oxide results when air is heated in furnaces, motor engine cylinders, or other high temperature combustion areas.

$$N_2(g) + O_2(g) \rightleftarrows 2\ NO(g)$$

a. Given that the standard enthalpy of formation for NO(g) is 90.25 kJ/mole and using the absolute entropies listed below, calculate the free energy change for this reaction at $25°C$. Is this reaction is product-favored at $25°C$?

Absolute Entropies

$N_2(g)$ 191.5 J/mol·K $O_2(g)$ 205.0 J/mole·K
NO(g)   210.7 J/mol·K

b. If the reaction is not product-favored at 25°C, determine under what conditions it would be product-favored.

6. (**L4**) Given that ΔG° = -100.4 kJ/mol for the reaction

$$CCl_4(\ell) \ + \ H_2(g) \ \rightleftharpoons \ HCl(g) \ + \ CHCl_3(g)$$

calculate the value of the equilibrium constant, K, for this reaction at 25°C.

7. When sulfur-containing fuels, such as coal or fuel oil, are burned, one of the products is SO_2, which is converted to SO_3 in the air according to the equation

$$2 \ SO_2(g) \ + \ O_2(g) \ \rightleftharpoons \ 2 \ SO_3(g)$$

Given the thermodynamic data in the table below, calculate K for this reaction at 25°C.

	ΔH_f° (kJ/mole)	S° (J/mole·K)
$SO_2(g)$	-296.8	248.2
$O_2(g)$	0	205.1
$SO_3(g)$	-395.7	256.8

PRACTICE PROBLEM SOLUTIONS

1. a. decrease, the solid is more ordered than the liquid.
   b. decrease, a less complex species is formed.
   c. increase, a solid dissolved in a liquid.
   d. decrease, a gas dissolving in a liquid.

2. Begin by writing the equation for the conversion of diamond to graphite.

$$C(\text{diamond}) \rightarrow C(\text{graphite})$$

Remembering that graphite is the standard state for carbon, calculate the free energy change for this process.

$$\Delta G^{\circ} = \Delta G^{\circ} (\text{graphite}) - \Delta G^{\circ} (\text{diamond})$$

$$= 0 - (2.90 \text{ kJ/mole})$$

$$\Delta G^{\circ} = -2.90 \text{ kJ/mole}$$

   The negative free energy change value indicates that diamond is spontaneously changing into graphite, but everyday experience suggests that diamonds seem to be quite permanent.
   This apparent contradiction is explained by the fact that thermodynamics tells us nothing about the <u>rate</u> of processes. Even though the diamond does tend to change into graphite, the rate of the transformation is extremely slow.

3. The equation that relates enthalpy of vaporization and entropy change is

$$\Delta S^{\circ} = \frac{H_{vap}}{T} = \frac{30.0 \times 10^{3} \text{ J/mol}}{350.0^{\circ}C}$$

$$\underline{\Delta S^{\circ} = 86 \text{ J/K} \cdot \text{mole}}$$

As would be expected, the entropy change is positive for the conversion of liquid molecules into gaseous molecules.

364

4. Substituting into the equation

$$\Delta G° = \Sigma \Delta G° \text{ (products)} - \Sigma \Delta G° \text{ (reactants)}$$

$$\Delta G°_f = 4\ \Delta G°_f\ (H_2O(g)) - \Delta G°_f\ (N_2H_4(\ell)) - 2\ \Delta G°_f\ (H_2O_2(\ell))$$

$$= 4(-228.6\ \text{kJ/mole}) - 149.2\ \text{kJ/mole}$$
$$- 2(-120.4\ \text{kJ/mole})$$

$$\underline{\Delta G° = -822.8\ \text{kJ}}$$

The large negative value agrees with the idea of using hydrazine as a rocket fuel, since it suggests that there is a great deal of energy that could be released to do work.

5. a. Determine the entropy change for this reaction.

$$\Delta S° = \Sigma S° \text{ (products)} - \Sigma S° \text{ (reactants)}$$
$$\Delta S° = 2\ S°(NO(g)) - S°(O_2(g)) - S°(N_2(g))$$
$$= 2(210.7\ \text{J/mol·K}) - (205.0\ \text{J/mol·K}) - (191.5\ \text{J/mol·K})$$
$$= 24.9\ \text{J/K}$$

Since two moles of NO are formed in the reaction

$$\Delta H° = 2(90.25\ \text{kJ/mol}) = 180.5\ \text{kJ}$$

Substituting these values into the equation

$$\Delta G° = \Delta H° - T\ \Delta S°$$

$$= 180.5\ \text{kJ} - (298\ \text{K})(24.9\ \text{J/K})(1 \times 10^{-3}\ \text{kJ/J})$$

$$\underline{\Delta G° = 173.1\ \text{kJ}}$$

The positive value indicates that this reaction will not be product-favored at 25°C.

b. As the temperature increases, the entropy term will have a greater tendency to make the reaction product-favored. To find the temperature at which the reaction will be at equilibrium, set $\Delta G°$ equal to zero.

$$\Delta G^o = 0 = \Delta H^o - T \Delta S^o$$

Now solve the equation for T

$$T = \frac{\Delta H^o}{\Delta S^o} = \frac{180.5 \text{ kJ}}{0.0249 \text{ kJ/K}}$$

$$\underline{T = 7250 \text{ K}}$$

<u>The reaction will be product-favored at temperatures above 7250 K.</u>

However, remember that even at lower temperatures the reaction may occur to a significant extent if there is little or no nitrogen oxide present initially.

6. The relationship between ΔG and K is given by the equation
$$\Delta G^o = - RT \ln K$$

Rearranging to solve for K

$$\ln K = \frac{-\Delta G^o}{RT} = \frac{-(-100.4 \text{ kJ/mol})(1 \times 10^3 \text{ J/kJ})}{(8.314 \text{ J/K} \cdot \text{mol})(298 \text{ K})}$$

$$\ln K = 40.5$$

$$\underline{K = 4.0 \times 10^{17}}$$

7. First, calculate the entropy change for the reaction.

$$\Delta S^o = 2 \ S^o(SO_3(g)) - 2 \ S^o(SO_2(g)) - S^o(O_2(g))$$

$$= 2(256.8 \text{ J/mol} \cdot \text{K}) - 2(248.1 \text{ J/mol} \cdot \text{K}) - (205.1 \text{ J/mol} \cdot \text{K})$$

$$\Delta S^o = -187.7 \text{ J/K}$$

Now, calculate the enthalpy change

$$\Delta H^o = 2 \ \Delta H^o_f \ (SO_3(g)) - 2 \ \Delta H^o_f \ (SO_2(g)) - \Delta H^o_f \ (O_2(g))$$

$$= 2(-395.7 \text{ kJ/mol}) - 2(-296.8 \text{ kJ/mol}) - (0)$$

$$\Delta H^\circ = -197.8 \text{ kJ}$$

Now, calculate the value of the free energy change

$$\Delta G^\circ = \Delta H^\circ - T \Delta S^\circ$$
$$= (-197.8 \times 10^3 \text{ J}) - 298 \text{ K}(-187.7 \text{ J/K})$$
$$\Delta G^\circ = -141,800 \text{ J}$$

Finally, use the free energy value to determine K

$$\Delta G^\circ = - RT \ln K$$

$$-141,800 \text{ J} = -(8.314 \text{ J/K·mol})(298 \text{ K}) \ln K$$

$$\ln K = 57.2$$

$$K = \underline{6.94 \times 10^{24}}$$

PRACTICE TEST (40 Minutes)

1. Without looking up any numerical values, predict whether the entropy will be increase or decrease for each of the following reactions:

a. $Br_2(\ell) \qquad \rightarrow \qquad Br_2(g)$

b. $2 \ Cl(g) \qquad \rightarrow \qquad Cl_2(g)$

c. $Ag^+(aq) + Cl^-(aq) \rightarrow AgCl(s)$

d. $H_2O(\ell) \qquad \rightarrow \qquad H_2O(s)$

e. $O_2(g) \qquad \rightarrow \qquad O_2(aq)$

2. Both entropy and enthalpy make a significant contribution to determining whether or not a reaction will be product-favored. Sometimes these two factors both have the same effect, but sometimes they work in opposite directions. In the latter case, temperature can be the deciding factor that determines reaction spontaneity. For each of the following cases, examine the sign of the enthalpy and entropy changes and suggest whether a process with those values would probably be

product-favored, reactant-favored, product-favored at low temperatures, or product-favored at high temperatures.

ΔH	ΔS	Prediction
+	+	
+	−	
−	+	
−	−	

3. The formation of photochemical smog is a complex process that involves hundreds of reactions. One of the important steps in this system is

$$O_3(g) \ + \ NO(g) \ \rightarrow \ NO_2(g) \ + \ O_2(g)$$

a. Calculate the value of ΔG° for this reaction using the thermodynamic values provided below.

b. Calculate the value of K for this reaction.

Standard Free Energies of Formation

$O_3(g)$ 163 kJ/mol   NO(g)   86.57 kJ/mol

$NO_2(g)$ 51.30 kJ/mol

4. Calculate the molar entropy of vaporization for liquid ammonia at its boiling point (−33°C). The enthalpy of vaporization for ammonia at the boiling point is 23. kJ/mol.

5. Nitrous oxide, N_2O, is used as a propellant for soft ice cream and as an anesthetic. Calculate the standard free energy of formation for this reaction and determine if it seems reasonable for the commercial production of this gas?

$$2 \ N_2O_4(g) \ \rightarrow \ 2 \ N_2O(g) \ + \ 3 \ O_2(g)$$

Standard Free Energies of Formation (kJ/mol)

$N_2O_4(\ell)$ 97.9 $N_2O(\ell)$ 104.2

6. At 25°C, the standard enthalpy of reaction is -91.3 kJ for the reaction

$$CCl_4(g) + H_2(g) \rightleftharpoons HCl(g) + CHCl_3(g)$$

Using the absolute entropy values provided, determine whether or not this reaction is product-favored at 25°C.

Absolute Entropy Values (J/mol·K)

$CCl_4(g)$ 309.7 $H_2(g)$ 130.6

$HCl(g)$ 186.8 $CHCl_3(g)$ 295.6

7. Using the standard free energies of formation provided below, calculate K at 25°C for the reaction

$$CO_2(g) + H_2(g) \rightleftharpoons CO(g) + H_2O(g)$$

Standard Free Energies of Formation (kJ/mole)

$CO(g)$ -137.2; $H_2O(g)$ -228.6; $CO_2(g)$ -394.4

COMMENTS ON ARMCHAIR EXERCISES

1. Of course, as you well know, salt does readily dissolve in water, so there must be some other factor besides the energy change that is involved. Consider the probable sign of the entropy change for this process; does this suggest a reason why salt dissolves in water?

2. Remember the section in the text that discussed the difference between thermodynamic control and kinetic control. Is it enough to know that a reaction is thermodynamically favored in order to predict whether it will actually occur to a significant extent during a reasonable time?

3. This is an obvious example where there is no energy difference between the randomly mixed marbles and those that are layered by color, but yet the mixed state is far more probable. What is

responsible for the difference in these states?

CONCEPT TEST ANSWERS

1. product-favored (or spontaneous)
2. no, it does not
3. a. having the energy dispersed over a greater number of atoms and molecules.
   b. having more disordered atoms or molecules
4. entropy
5. zero, third law of thermodynamics
6. A,b, and d represent an increase in entropy; c represents a decrease in entropy.
7. entropy
8. product-favored
9. negative
10. standard molar free energy of formation
11. zero
12. 8.314 J/K·mol

PRACTICE TEST ANSWERS

1. (L1) a. increase, liquid converted to a gas
   b. increase, a more complex species forms
   c. decrease, a precipitate forms
   d. decrease, liquid changes to a solid
   e. decrease, gas dissolves in a solvent

2. (L3)

ΔH	ΔS	Prediction
+	+	product-favored at high temperatures
+	−	reactant-favored
−	+	product-favored
−	−	product-favored at low temperatures

3. a. (L2) −198.0 kJ
   b. (L4) K = 5.6×10^{34}

4. (L1) 96 J/K·mole

5. (L2) ΔG° = 12.6 kJ. The positive free energy for this reaction suggests that it would not be useful, at least under these conditions.

6. **(L3)** $\Delta S^o = 42.0$ J/K; $\Delta G^o = -105.0$ kJ; The reaction is product-favored

7. **(L4)** $\Delta G = 28.6$ kJ; $K = 1.03 \times 10^{-5}$

CHAPTER 21
PRINCIPLES OF REACTIVITY:
ELECTRON TRANSFER REACTIONS

CHAPTER OVERVIEW

Even though it may not be apparent, electrochemistry is vital for many everyday activities. Batteries are used to generate a usable electric current for many portable devices, ranging from TVs to toys. They also are used for automobiles and other forms of transportation. They provide a source of electricity, even when there is no permanent connection available, and so help to make our lives more convenient.

Just as chemical reactions can be used to create an electric current, an electric current can be used to create a chemical reaction. This process, called electrolysis, is very useful to the chemist but it also allows many different items to be plated with a thin coat of metal, allowing it to be preserved or to acquire an attractive finish at a relatively low cost.

The third major area, corrosion, is less beneficial but equally important. Tarnishing of silver, rusting of cars and metal appliances, and many other forms of corrosion are simply the result of electrochemistry. If the process is understood, it is frequently possible to control or even eliminate it.

LEARNING GOALS

1. You should be able to balance equations for oxidation-reduction reactions in acidic and basic solution. (Sec. 21.1)

2. Understand how an oxidation-reduction reaction in a voltaic cell can be used to produce an electric current, recognize the various components of such a cell (anode, cathode, and salt bridge). (Sec. 21.2)

3. Since ΔG is negative for product-favored

reactions, the equation
$$\Delta G = -nFE^\circ$$

indicates that all product-favored electron transfer reactions have a positive value for E°. Be able to use a table of standard reduction potentials (such as Table 21.1 in the textbook) to calculate the E° value for the combination of a pair of half-reactions and use this result to predict whether or not this combination will occur spontaneously under standard conditions. (Sec. 21.3 and 21.4)

4. When the conditions in an electrochemical cell are nonstandard, be able to use the Nernst equation,

$$E = E^\circ - \frac{2.303RT}{nF} \log Q$$

to calculate the potential of an electrochemical cell. (Sec. 21.5)

5. Also be able to use the Nernst equation to determine equilibrium constants from standard reduction potential values. (Sec. 21.5)

6. A variety of different voltaic cells are commonly used to provide electric current, including the dry cell, the storage battery, the mercury battery, the alkaline battery, and fuel cells. Be familiar with the chemistry involved in some of these systems. (Sec. 21.6)

7. Corrosion reactions are an extremely important type of oxidation-reduction process. Understand the conditions that are most likely to produce corrosion as well as how corrosion can be prevented. (Sec. 21.7)

8. Electrolysis is an important commercial technique, and so it is important to be able to predict the products that result when an electric current is passed through various types of solutions or melted solids. Be able to use standard reduction potentials to predict the most likely products of electrolysis reactions and also be familiar with some of the important commercial applications of electrolysis. (Sec. 21.8 and 21.10)

9. The Faraday constant serves as the basis for a useful relationship between current flow and the amount of chemical reaction that can occur. It's important to be able to do these conversions in either direction and also to be able to determine the amount of energy involved in units of either joules or kilowatt-hours. (Sec. 21.9)

ARMCHAIR EXERCISES

1. You have probably seen many different sizes of batteries, ranging from the storage battery in your car to the small batteries used in hearing aids. Does the size of the battery affect the voltage it will deliver?

2. A simple way to create an electrochemical cell is to cut a slice of lemon and push a copper penny and an iron nail into the meat of the lemon. If the probes of a voltmeter are connected to the two metal items, a current is observed to flow.
    Would the voltage be significantly different if you had used an orange instead of a lemon? What purpose does the lemon serve? What is primarily responsible for determining the voltage? If you connect the two pieces of metal for some time, you may well find that the iron nail has changed. How would you expect it to change? If you actually try this experiment, THROW THE LEMON AWAY AND DON'T EAT IT! Why is this?

3. If you or someone you know has mercury amalgam dental fillings, you may be familiar with a rather unpleasant aspect of electrochemistry. If the aluminum wrapper from a piece of candy comes in contact with the filling, an unpleasant jolt of pain results. Can you suggest an explanation for this problem?

CONCEPT TEST

1. In an electrochemical cell, oxidation always occurs at the electrode called the _____, and reduction always occurs at the _____.

2. In order to maintain a balance of ion charges in the compartments of an electrochemical cell, the two half-cells are often connected with a device called a(n) _____.

3. The value of E^0 for all product-favored electron transfer reactions must be _____.

4. Using Table 21.1, arrange the following ions in order of increasing strength as oxidizing agents: Ag^+, Na^+, and Sn^{2+}.

(weakest) _____ < _____ < _____ (strongest)

5. Unless specified otherwise, all values of E^0 are given at a temperature of _____.

6. The standard cell potentials for all half-cell reactions is measured relative to the _____ _____ electrode.

7. Determine the value of n (as used in the Nernst equation) for each equation. (Be sure the equation is balanced first!)

a. Cu^{2+} + Cd → Cd^{2+} + Cu

b. Al + Sn^{2+} → Al^{3+} + Sn

c. Na + Cl_2 → Cl^- + Na^+

8. When the cell potential, E, has a value of _____, the Q term in the Nernst equation is equivalent to the equilibrium constant, K.

9. Sodium metal is produced in a Downs cell by the electrolysis of _____.

10. Chlorine is produced by the electrolysis of _____.

11. Protecting a metal object from corrosion by painting the surface or forming an oxide coating is called _____.

12. If a substance has a reduction potential more negative than _____ volts, then only water will be reduced when we pass an electric current through an aqueous solution of this species.

13. In a battery, the polarity of the anode is ___, but for an electrolysis cell, the polarity of the anode is ___. (Choose from + or - in each case.)

14. In many cases, the current measured when a cell is working is found to be different from that calculated, even though the cell components are at standard conditions. This difference is called _____, and is primarily due to the effect of reaction kinetics.

STUDY HINTS

1. It is essential that you follow a systematic procedure when balancing oxidation-reduction equations. If you learn each step of the method in your textbook and conscientiously do each step for every oxidation-reduction equation you balance, you will probably find that most of your errors are due to a lapse of concentration. Some common errors of this type are (a) forgetting to make sure that other elements are balanced before balancing hydrogen and oxygen, (b) multiplying only the reactants in a half reaction by the correct factor, or (c) forgetting to simplify the final equation.

The charges on the ions are very important, so be sure to write the charges clearly and neatly. Otherwise it's very easy to overlook an ion in adding up the charges and the balancing will be incorrect.

2. Remember that in the Nernst equation, the value of n is determined by the number of electrons transferred when balancing the net equation. Normally is isn't possible to simply look at one half reaction and determine what the n value will be for a net equation. Be sure to balance the two

half reactions in the usual way, and then determine n from the number of electrons that are canceled out when the two half reactions are added.

3. In many cases, chemists have agreed to always do certain things the same way when writing electrochemistry problems. For example, the anode is normally written on the left of an electrochemical cell, and the table of standard electrode potentials is normally written with the most negative potentials on the top of the list. Don't depend too much on these conventions. It can be a rude shock for those who memorize that the strongest oxidizing agents are at the top of the reduction potential table, then encounter a table that lists the half reactions in the opposite order. It's always better to try to understand rather than just memorize isolated facts, but that is especially true in electrochemistry.

4. When using the table of standard reduction potentials, don't forget that the half-reactions listed include both an oxidizing agent and a reducing agent. When asked to identify the oxidizing agent in a process, if you give the complete half-reaction, the answer contains both an oxidizing agent and a reducing agent and may be marked wrong for that reason.

PRACTICE PROBLEMS

1. (L6) Briefly describe the following common batteries, including identifying the anode and cathode. (a) dry cell (b) mercury cell

2. **(L2)** Using Table 21.1 in your text, draw a simple diagram of each voltaic cell below. Assume the metal ions are 1 Molar aqueous solutions and use an inert platinum electrode if necessary. Label the cells thoroughly, including the anode and cathode, the direction that the electrons move in the circuit, and assuming that the salt bridge contains $NaNO_3$, the direction the NO_3^- ions move.

A. $Cd(s) \mid Cd^{2+}(1\ M) \parallel Sn^{2+}(1\ M) \mid Sn$

B. $Fe(s) \mid Fe^{2+}(1\ M) \parallel Cu^{2+}(1\ M) \mid Cu$

3. **(L3)** Use the following table of reduction potentials to answer the various questions below.

Half Reaction	$E^0(V)$
$Cl_2(g)\ +\ 2\ e^- \rightarrow 2\ Cl^-(aq)$	+1.360
$Ag^+(aq)\ +\ e^- \rightarrow Ag(s)$	+0.80
$Cu^{2+}(aq)+\ 2\ e^- \rightarrow Cu(s)$	+0.337
$Ni^{2+}(aq)+\ 2\ e^- \rightarrow Ni(s)$	-0.25
$Cd^{2+}(aq)+\ 2\ e^- \rightarrow Cd(s)$	-0.40
$Mg^{2+}(aq)+\ 2\ e^- \rightarrow Mg(s)$	-2.37

Based only on the above table:

a. _____ is the strongest oxidizing agent in the above table.

b. _____ is the strongest reducing agent in the above table.

c. _____ are the symbols of the metals

378

from the table that will reduce copper(II) ion.
d. _____ are the symbols of metals from
the table can be oxidized by silver(I) ion.
e. Will chlorine gas oxidize magnesium metal to
magnesium(II) ion?   _____

4. (L9) A current of 2.50 amperes was passed
through a solution of chromium(III) chloride for
3.50 hours.   What mass of chromium metal was
deposited on the cathode of the electrolytic cell?

5. (L4) Using the table of reduction potentials in
your text, calculate the voltage for the following
electrochemical cell:

$$Cd(s) \mid Cd^{2+}(0.20 \text{ M}) \mid\mid Sn^{2+}(0.80 \text{ M}) \mid Sn(s)$$

6. (L5) The standard reduction potential equals
-1.245 volts in basic solution for the half
reaction

$$Zn(OH)_2(s) + 2 e^- \rightarrow Zn(s) + 2 OH^-(aq)$$

Using Table 21.1 in the text, select an appropriate half reaction and determine the K_{sp} value for $Zn(OH)_2$.

7. **(L8)** As described in your textbook, aluminum is prepared by the Hall-Heroult process, which involves the electrolysis of a molten mixture of aluminum oxide and cryolite, Na_3AlF_6. a. How many coulombs of electricity would be required to produce 1000 kilograms of aluminum metal? b. If the electrolysis cell operates at 5.0 volts, how many kilowatt-hours of electricity will be required to produce this much aluminum?

8. **(L1)** Balance the following oxidation-reduction equations.

a. $Mn + Cr^{3+} \rightarrow Mn^{2+} + Cr$

b. $Sn^{2+} + Cr_2O_7^{2-} \rightarrow Cr^{3+} + Sn^{4+}$ (acid solution)

c. $AsO_2^- + ClO^- \rightarrow AsO_3^- + Cl^-$ (basic solution)

PRACTICE PROBLEM SOLUTIONS

1. The common dry cell contains a carbon rod electrode inserted into a moist paste of NH_4Cl, $ZnCl_2$, and MnO_2 in a zinc can that serves as the anode.

The mercury battery, which is used in many devices that require a small cell, uses metallic zinc as the anode and mercury (II) oxide as the cathode.

2.

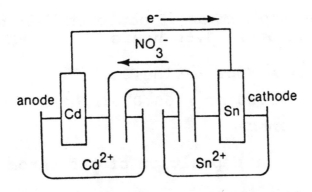

Figure 21.1 Answer to Practice Problem 2a.

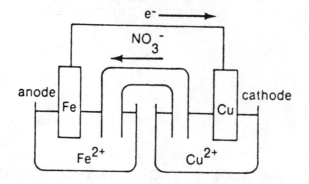

Figure 21.2 Answer to Practice Problem 2b.

3. a. $Cl_2(g)$ is the strongest oxidizing agent
b. $Mg(s)$ is the strongest reducing agent
c. Magnesium, cadmium, and nickel will reduce copper(II).
d. Copper, nickel, cadmium, and magnesium are oxidized by silver(I).
e. The E^o value for the reaction would be +3.73 volts, so the reaction is very product-favored.

4. Since you are asked to find the grams of chromium, the calculations should be planned to obtain the moles of chromium. This can be calculated if you know the moles of electrons that have been transferred. That will depend on the current and the time. This sequence of calculations can be represented as follows:

Time x current → charge → moles → moles
of electrons of Chromium

First, determine the number of coulombs that passed through the solution in 3.50 hours.

Charge (coulombs) = amps x time(sec)

   = 2.50 amps x 3.50 hr x 60.0 min/hr x 60 s/min

   = 31,500 coulombs

The half reaction for deposition of the chromium is

$$Cr^{3+}(aq) \ + \ 3e^- \ \rightarrow \ Cr(s)$$

382

Therefore the deposition requires three moles of electrons per mole of chromium deposited.

The number of moles of electrons is then

$$\text{Faradays} = \frac{(31,500 \text{ coulombs})}{(96,500 \text{ coulombs/Faraday})} \times \frac{1 \text{ mole Cr}}{3 \text{ Faradays}}$$

Moles of Chromium = 0.109 moles

Now determine the mass of chromium produced.

Mass of Cr = 0.109 moles x 52.00 g/mole

Mass of Cr = 5.66 grams

5. From Table 21.1, the voltages of the half reactions are:

Cd^{2+} + 2e⁻ → Cd(s) E^o = -0.40 Volts

Sn^{2+} + 2e⁻ → Sn(s)   = -0.14

Reverse the cadmium half reaction, and add the half reactions to produce the overall reaction:
Oxidation

Cd(s) → Cd^{2+} + 2e⁻ E^o = +0.40 V

Reduction

Sn^{2+} + 2e⁻ → Sn(s) E^o = -0.14 V
_____

Sn^{2+} + Cd(s) → Cd^{2+} + Sn(s) E^o_{cell} = +0.26 V

Since the concentrations of the reactants are not 1 M, it is necessary to use the Nernst equation to determine the effect of the nonstandard concentrations.

$$E = E^o - \frac{2.303RT}{nF} \log Q$$

$$E = E^o - \frac{0.0592}{n} \log \frac{[Cd^{2+}]}{[Sn^{2+}]}$$

$$= 0.26 \text{ V} - \frac{0.0592}{2 \text{ mol electrons}} \log \frac{0.20\text{M}}{0.80\text{M}}$$

$$E = 0.26 \text{ V} + 0.0178 \text{ V}$$

$$\underline{E_{cell} = 0.28 \text{ V}}$$

The result of the concentration change was to make the cell reaction more product-favored.

6. If you combine the half reaction provided with the half reaction for zinc ion, you will obtain the solubility product reaction for $Zn(OH)_2$:

$$Zn(OH)_2(s) + 2 \text{ e}^- \rightarrow Zn(s) + 2 \text{ OH}^-(aq) \quad E^\circ = -1.245 \text{ V}$$

$$\underline{Zn(s) \quad \rightarrow \quad Zn^{2+} + 2 \text{ e}^- \qquad\qquad\qquad = \quad 0.763 \text{ V}}$$

$$Zn(OH)_2(s) \rightarrow Zn^{2+}(aq) + 2 \text{ OH}^-(aq) \quad E^\circ = -0.482 \text{ V}$$

Once you have determined the value of E° for the reaction, it is easy to substitute into the Nernst equation and solve for K:

$$E = E^\circ - \frac{2.303RT}{nF} \log K$$

Setting E = 0 at equilibrium

$$E^\circ = \frac{2.303RT}{nF} \log K$$

$$\log K = \frac{(2.0 \text{ mole})(-0.482 \text{ V})}{0.0592} = \frac{- 0.964 \text{ V·mole}}{0.0592 \text{ V·mole}}$$

$$\log K = - 16.28$$

$$\underline{K = 5.2 \times 10^{-17}}$$

7. First, calculate the moles of aluminum to be formed.

$$\text{Moles of aluminum} = \frac{1000 \text{ kg} \times 1 \times 10^3 \text{ g/kg}}{27.0 \text{ g/mole}}$$

Moles of aluminum = 3.7×10^4 moles

The half reaction for the deposition of aluminum is

$$Al^{3+}(\text{melt}) + 3 e^- \rightarrow Al(s)$$

3 moles of electrons will produce 1 mole of Al.

$$\text{mole of } e^- = \frac{(3.7 \times 10^4 \text{ moles of Al})(3 \text{ moles of } e^-)}{(1 \text{ mole Al})}$$

moles of e^- required = 1.11×10^5 moles of e^-

Next determine the coulombs needed

$$\text{coulombs} = (1.11 \times 10^5 \text{ moles of } e^-)\frac{(96,500 \text{ coulombs})}{\text{mole of electrons}}$$

coulombs = 1.1×10^{10} coulombs

7. b. To calculate the number of kilowatt-hours, first determine the number of joules

Joules = coulombs x volts

$\quad\quad = (1.1 \times 10^{10} \text{ coulombs})(5.0 \text{ V})$

Joules = 5.4×10^{10} joules

Now calculate the number of kilowatt-hours.

kwh = $(5.4 \times 10^{10} \text{ joules})(1 \text{ kwh}/3.6 \times 10^6 \text{ J})$

kilowatt-hours = 1.5×10^4 kwh

8. a. Since the question indicates that these are oxidation-reduction reactions, proceed to Step 2. and write the half reactions.

$$Mn \rightarrow Mn^{2+} \quad \text{(oxidation)}$$

$$Cr^{3+} \rightarrow Cr \quad \text{(reduction)}$$

<u>Step 3.</u> Balance the mass, starting with elements other than hydrogen and oxygen. Both equations are balanced as written.

$$Mn \rightarrow Mn^{2+}$$

$$Cr^{3+} \rightarrow Cr$$

<u>Step 4.</u> Balance the charge by adding electrons to the side of each equation that has the highest positive charge.

$$Mn \rightarrow Mn^{2+} + 2e^-$$

$$Cr^{3+} + 3e^- \rightarrow Cr$$

Notice that the electrons must appear on opposite sides of the two equations. Can you see why this is essential?

<u>Step 5.</u> The manganese reaction produces two electrons each time it occurs, but the chromium reaction requires three electrons each time it occurs. To make the number of electrons produced equal the number required, multiply the manganese equation by three and the chromium reaction by two.

$$3(Mn \rightarrow Mn^{2+} + 2e^-)$$

$$2(Cr^{3+} + 3e^- \rightarrow Cr)$$

Don't forget to multiply the entire equation!

<u>Step 6.</u> Now add the two half reactions together.

$$3 Mn \rightarrow 3 Mn^{2+} + 6 e^-$$

$$2 Cr^{3+} + 6 e^- \rightarrow 2 Cr$$

$$\overline{3 Mn + 2 Cr^{3+} \rightarrow 3 Mn^{2+} + 2 Cr}$$

<u>Step 7.</u> The mass and charge are balanced on both sides of the equation. In addition, the coefficients cannot be simplified, and none of the species can be canceled as spectator ions. Thus the equation is balanced.

386

13b. <u>Step 2.</u> The two half reactions are

$$Sn^{2+} \rightarrow Sn^{4+}$$

$$Cr_2O_7^{2-} \rightarrow Cr^{3+}$$

<u>Step 3.</u> Begin by balancing everything <u>except</u> H and O. The other elements are often balanced without changing any coefficients, but always check to make sure. In this case, the chromium is not balanced in the second half reaction.

$$Sn^{2+} \rightarrow Sn^{4+}$$

$$Cr_2O_7^{2-} \rightarrow 2\ Cr^{3+}$$

The tin half reaction is now balanced for mass, so it will not be necessary to do anything further with it until the next step.

Next, balance oxygen. Since the solution is acidic hydrogen ion and water can be added as needed. The reactant side of the chromium half reaction has seven oxygens, the product side none. To balance oxygen, add seven water molecules to the product side.

$$Cr_2O_7^{2-} \rightarrow 2\ Cr^{3+} + 7\ H_2O$$

The last step in balancing the mass is to balance hydrogen by adding hydrogen ions. There are 14 hydrogens on the product side, but none on the reactant side. Therefore add 14 hydrogen ions to the reactant side.

$$14\ H^+ + Cr_2O_7^{2-} \rightarrow 2\ Cr^{3+} + 7\ H_2O$$

<u>Step 4.</u> Balance the charge, adding electrons to the more positive side of each half reaction until both sides have the same charge.

$$Sn^{2+} \rightarrow Sn^{4+} + 2\ e^-$$

$$6\ e^- + 14\ H^+ + Cr_2O_7^{2-} \rightarrow 2\ Cr^{3+} + 7\ H_2O$$

<u>Step 5.</u> Next, multiply by appropriate factors so that the same number of electrons are donated and

consumed.   The simplest way to do this is to multiply the tin half reaction by 3, so that 6 moles of electrons are donated and six moles are consumed.

$$3 \; (Sn^{2+} \quad \rightarrow \quad Sn^{4+} \quad + \quad 2 \; e^-)$$

$$6 \; e^- + 14 \; H^+ + Cr_2O_7^{2-} \quad \rightarrow \quad 2 \; Cr^{3+} \quad + \quad 7 \; H_2O$$

Step 6. Now the half reactions are added together to give the overall equation.

$$3 \; (Sn^{2+} \quad \rightarrow \quad Sn^{4+} \quad + \quad 2 \; e^-)$$

$$6 \; e^- + 14 \; H^+ + Cr_2O_7^{2-} \quad \rightarrow \quad 2 \; Cr^{3+} \quad + \quad 7 \; H_2O$$

$$14 \; H^+ + Cr_2O_7^{2-} + 3 \; Sn^{2+} \rightarrow 3 \; Sn^{4+} + 2 \; Cr^{3+} + 7 \; H_2O$$

Step 7. Notice that the total charge on the reactant side is 18+, and the total charge on the product side is also 18+. Also each side has 14 H, 7 O, 2 Cr, and 3 Sn, so the mass is also balanced.

There are no species on both sides of the equation that can be canceled out, and you cannot divide the coefficients by a constant factor. So the simplest, balanced, ionic equation is

$$14 \; H^+ + Cr_2O_7^{2-} + 3 \; Sn^{2+} \rightarrow 3 \; Sn^{4+} + 2 \; Cr^{3+} + 7 \; H_2O$$

13c.   Step 2. Write the half reactions

$$AsO_2^- \quad \rightarrow \quad AsO_3^-$$

$$ClO^- \quad \rightarrow \quad Cl^-$$

Step 3. Balance each half reaction for mass

Arsenic and chlorine are balanced, so move directly to balance oxygen. Count the number of oxygens on each side of the reaction, then add water to the side that has the most oxygens, using one water for each excess oxygen.

$$AsO_2^- \quad \rightarrow \quad AsO_3^- \quad + \quad H_2O$$

$$H_2O + ClO^- \quad \rightarrow \quad Cl^-$$

Next, on the opposite side from that where you

added the water, add two hydroxides for each water molecule that you added.

$$2 \, OH^- + AsO_2^- \rightarrow AsO_3^- + H_2O$$

$$H_2O + ClO^- \rightarrow Cl^- + 2 \, OH^-$$

Step 4. Now balance the half reactions for charge.

$$2 \, OH^- + AsO_2^- \rightarrow AsO_3^- + H_2O + 2 \, e^-$$

$$2 \, e^- + H_2O + ClO^- \rightarrow Cl^- + 2 \, OH^-$$

Step 5. The number of electrons lost in the first equation equals the number gained in the second, so it isn't necessary to multiply by a factor.

$$2 \, OH^- + AsO_2^- \rightarrow AsO_3^- + H_2O + 2 \, e^-$$

$$2 \, e^- + H_2O + ClO^- \rightarrow Cl^- + 2 \, OH^-$$

Step 6. Add up the half reactions to give the overall reaction.

$$2 \, OH^- + 2 \, AsO_2^- \rightarrow 2 \, AsO_3^- + H_2O + 2 \, e^-$$

$$2 \, e^- + H_2O + ClO^- \rightarrow Cl^- + 2 \, OH^-$$

$$2 \, OH^- + 2 \, AsO_2^- + H_2O + ClO^- \rightarrow$$
$$Cl^- + 2 \, OH^- + 2 \, AsO_3^- + H_2O$$

Notice that hydroxide and water appear on both sides of the equation. We can simplify the equation by canceling species that appear on both sides of the equation. This means that all of the hydroxides and both of the waters will cancel out, leaving the simplified equation as follows.

$$AsO_2^- + ClO^- \rightarrow Cl^- + AsO_3^-$$

PRACTICE TEST (45 min.)

1. Using Table 21.1 in your text, draw a simple diagram of the voltaic cell given below. Use an inert platinum electrode if necessary and assume

that the dissolved ions are in 1 Molar aqueous solution. Label the cell thoroughly, including the anode and cathode, the direction of motion of the electrons in the circuit, and assuming that the salt bridge contains $NaNO_3$, the direction of movement of the NO_3^- ions.

A. $Al(s)$ + Zn^{2+} $\rightarrow$ Al^{3+} + $Zn(s)$

2. First, balance each equation below and then use Table 21.1 in your textbook to calculate the standard potential, E^o, and predict which of these reactions will occur spontaneously in the direction in which they are written.

a. $Al(s)$ + $Cu^{2+}(aq)$ $\rightarrow$ $Al^{3+}(aq)$ + $Cu(s)$

b. $Zn(s)$ + $Mg^{2+}(aq)$ $\rightarrow$ $Zn^{2+}(aq)$ + $Mg(s)$

c. $Cd(s)$ + $Ag^+(aq)$ $\rightarrow$ $Cd^{2+}(aq)$ + $Ag(s)$

d. $Hg^{2+}(aq)$ + $PbSO_4(s)$ + $H_2O(\ell)$
$\rightarrow$ $Hg(\ell)$ + $PbO_2(s)$ + $SO_4^{2-}(aq)$ + $H^+(aq)$

3. A current of 2.00 amps is passed through a solution of copper nitrate until 6.35 grams of copper metal has been deposited.   a. How many seconds did the current flow?   b. How many hours did the current flow?

4. An electrochemical concentration cell is set up as shown:

$Cu(s) \mid Cu^{2+}(0.10 \text{ M}) \parallel Cu^{2+}(10.0 \text{ M}) \mid Cu(s)$

What is the initial voltage of this cell when it is operated with the concentrations indicated?

5. Determine the cell voltage for the following cell

$Zn(s) \mid Zn^{2+}(0.10 \text{ M}) \parallel Ag^+(0.0010 \text{ M}) \mid Ag(s)$

if the voltage for this cell under standard conditions, E^o, is 1.56 volts.

6. The standard reduction potential equals 0.222 volts for the half reaction

$$AgCl(s) + e^- \rightarrow Ag(s) + Cl^-(aq)$$

Using Table 21.1 in the text, select an appropriate half reaction and determine the K_{sp} value for AgCl.

7. Indicate the chemical reaction at the anode and the cathode if an electric current is passed through each of the following:

a. $MgCl_2$(aqueous)
b. $MgCl_2$(molten)
c. LiBr(aqueous)
d. $CdCl_2$(aqueous)

8. Complete and balance this oxidation-reduction equation in acidic solution.

$$C_2O_4^{2-} + Cr_2O_7^{2-} \rightarrow Cr^{3+} + CO_2 + H_2O$$

COMMENTS ON ARMCHAIR EXERCISES

1. The size of the battery doesn't affect the voltage; that is determined by the composition of the two electrodes. What effect is related to the size of the battery?

2. The voltage is created by the difference between the two metals, so changing to an orange or grapefruit should make little difference in the voltage observed. You may well have seen a product called a potato clock, where two metal electrodes are placed in a potato and provide enough voltage to run a small clock. This is the same idea. As the reaction continues what do you expect to happen at the iron nail? What do you expect to happen at the copper electrode? Does this explain why you should not eat the lemon?

3. When the mercury comes in contact with the filling, an electrochemical cell is formed with the mercury as the cathode and the aluminum as the anode. The nerve in the tooth is sensitive enough to detect this current flow.

CONCEPT TEST ANSWERS

1. anode, cathode
2. salt bridge
3. a positive number
4. (weakest) $Na^+ < Sn^{2+} < Ag^+$ (strongest)
5. $25°C$
6. standard hydrogen electrode (or S.H.E.)
7. a. two   b. six   c. two
8. zero
9. molten sodium chloride
10. aqueous sodium chloride solution
11. anodic inhibition
12. about -0.8 volts
13. negative (-), positive (+)
14. overvoltage

PRACTICE TEST ANSWERS

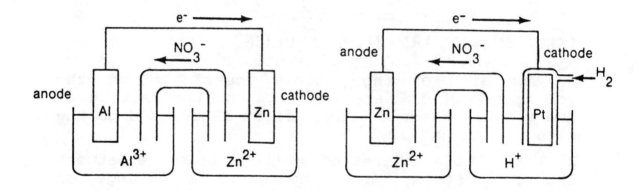

Figure 21.3 Answer to practice test, question 1.

2. (L3)

a. $2 Al(s) + 3 Cu^{2+}(aq) \rightarrow 2 Al^{3+}(aq) + 3 Cu(s)$
    $E° = +1.997$ V, product-favored

b. $Mg^{2+}(aq) + Zn(s) \rightarrow Mg(s) + Zn^{2+}(aq)$
    $E° = -1.607$ V, reactant-favored

c. $Cd(s) + 2 Ag^+(aq) \rightarrow Cd^{2+}(aq) + 2 Ag(s)$
    $E° = +1.20$ V, product-favored

d. $Hg^{2+}(aq) + PbSO_4(s) + 2 H_2O(\ell)$
    $\rightarrow Hg(\ell) + PbO_2(s) + SO_4^{2-} + 4 H^+$
    $E° = -0.830$ V, reactant-favored

3. **(L9)** a. 9640 seconds, b. 2.68 hours
4. **(L4)** 0.0592 V
5. **(L4)** voltage = 1.42 volts
6. **(L5)** 1.7×10^{-10}
7. a. Anode (oxidation of chloride)

$$2 \ Cl^-(aq) \ \rightarrow \ Cl_2(g) \ + \ 2 \ e^-$$
Cathode (reduction of water)

$$2 \ H_2O(\ell) \ + \ 2 \ e^- \ \rightarrow \ H_2(g) \ + \ 2 \ OH^-(aq)$$
b. Anode (oxidation of chloride)

$$2 \ Cl^-(aq) \ \rightarrow \ Cl_2(g) \ + \ 2 \ e^-$$
Cathode (reduction of magnesium)

$$Mg^{2+}(aq) \ + \ 2 \ e^- \ \rightarrow \ Mg(s)$$
c. Anode (oxidation of bromide)

$$2 \ Br^-(aq) \ \rightarrow \ Br_2(g) \ + \ 2 \ e^-$$
Cathode (reduction of water)

$$2 \ H_2O(\ell) \ + \ 2 \ e^- \ \rightarrow \ H_2(g) \ + \ 2 \ OH^-(aq)$$
d. Anode (oxidation of chloride)

$$2 \ Cl^-(aq) \ \rightarrow \ Cl_2(g) \ + \ 2 \ e^-$$
Cathode (reduction of cadmium)

$$Cd^{2+}(aq) \ + \ 2 \ e^- \ \rightarrow \ Cd(s)$$
8. **(L1)** $3 \ C_2O_4^{2-} \ + \ Cr_2O_7^{2-} \ + \ 14 \ H^+$

$$\rightarrow \ 2 \ Cr^{3+} \ + \ 6 \ CO_2 \ + \ 7 \ H_2O$$

CHAPTER 22
THE CHEMISTRY OF THE
MAIN GROUP ELEMENTS

CHAPTER OVERVIEW

This chapter provides an opportunity to review much of the chemistry that you have learned during the course. The number of different reactions is a little frightening, but remember to use the periodic table to help you to organize the reactions as much as possible. It's also helpful if you stop after reading about each element and think about where you might encounter products related to that element during your everyday life.

LEARNING GOALS

1. You should review what you have learned about the periodic table, especially the relationship between electronic configuration and bonding. (Sec. 22.1)

2. Learn about the chemical and physical properties, methods of preparation, compounds, and uses of hydrogen. (Sec. 22.2)

3. Learn about the properties, methods of preparation, common compounds, and uses of sodium and potassium. (Sec. 22.3)

4. Learn about the properties, methods of preparation, common compounds, and uses of calcium and magnesium. (Sec. 22.4)

5. Learn about the properties, methods of preparation, common compounds, and uses of aluminum and silicon. (Sec. 22.5 and 22.6)

6. Learn about the chemical and physical properties, methods of preparation, compounds, and uses of nitrogen and phosphorus. (Sec. 22.7)

7. Learn about the chemical and physical properties, methods of preparation, compounds, and uses of oxygen and sulfur. (Sec. 22.8)

8.   Learn about the chemical and physical properties, methods of preparation, compounds, and uses of chlorine. (Sec. 22.9)

CONCEPT TEST

1. Water gas is a mixture of carbon monoxide and _____.

2.   Hydrogen will form chemical combinations with every element except those in the periodic group called the _____.

3. The largest commercial use of hydrogen gas is the Haber process, which is a major source of _____ gas.

4.   When hydrogen reacts with electronegative elements, such as carbon, nitrogen, oxygen, and fluorine, so form covalent hydrides, the formal oxidation number of the hydrogen is _____.

5. Sodium and potassium are quickly coated with an oxide film when exposed to air, and so they are normally stored under _____.

6. The principal source of magnesium metal is _____.

7. Match each chemical formula on the left with its common name on the right.

_____ $NaHCO_3$	A. lime
_____ CaO	B. soda ash
_____ Na_2CO_3	C. baking soda

8. Water that contains significant amounts of the ions _____ and _____ is called hard water.

9. The chemistry of the alkali metals is dominated by the _____ oxidation state; the chemistry of the alkaline earth metals by the _____ oxidation state.

10. When lime mortar is placed between bricks or stone blocks, it slowly absorbs carbon dioxide from the air and becomes _____.

11. Aluminum is obtained by the electrolysis of hydrated aluminum oxide, called _____.

12. Although aluminum is widely used as a structural material and for packaging, the pure element is rarely used. Why?

_____

13. Aluminum is highly corrosion resistant because a thin film of _____ forms rapidly on the surface of aluminum metal.

14. _____ is the second most abundant element in the earth's crust.

15. Quartz is a pure crystalline form of the simplest oxide of silicon, which is called

_____.

16. Match each of the nitrogen oxides on the left below with the best descriptive phase on the right. Use each phrase on the right only once.

_____ NO   a. laughing gas

_____ N_2O_4 b. brown air pollutant

_____ N_2O c. product when NO_2 dimerizes

_____ NO_2 d. intermediate in the production of nitric acid by oxidation of ammonia

17. _____ is by far the most abundant element in the earth's crust.

18. _____ is the blue, diamagnetic gas that is the primary allotrope of oxygen.

19. _____ is the compound

produced in the largest quantity by the chemical industry in the United States.

20. _____ is the element that normally exists as a highly-reactive, yellow-green gas.

STUDY HINTS

1. It will be easier to remember the equations in the next few chapters if general types of reactions are identified and learned together rather than just memorizing each equation separately. For instance, try to pick out examples of the following cases in this chapter:

(a) direct combination of one element with another,

(b) reduction with a reducing agent, such as carbon monoxide or hydrogen, and

(c) replacement of a less active metal with a more active metal.

PRACTICE PROBLEMS

1. (All Sections) Complete and balance the following equations:

a. $H_2(g)$ + $F_2(g)$ →

b. $H_2O(g)$ + $CO(g)$ $\xrightarrow{\text{heat}}$

c. $2 Na(s)$ + $H_2(g)$ →

d. $Na(g)$ + $KCl(\ell)$ →

e. $2 Na(s)$ + $Cl_2(g)$ →

f. $2 Na(s)$ + $O_2(g)$ →

g. $SiO_2(s)$ + $3 C(s)$ $\xrightarrow{3000°C}$

h. $2 KClO_3(s)$ →

i. $Cl_2(g)$ + $2 OH^-(\text{cold aq})$ →

2. **(L1)** List at least three elements that will liberate hydrogen gas from dilute acidic solution. (Hint: Use the table of reduction potentials in Chapter 21.)

3. **(L1, L4, & L6)** Write balanced chemical reactions for one method of preparation for each substance listed.
a. ammonia

b. overall reaction for the Solvay process forming soda ash

c. hydrogen chloride

4. **(L2)** You are already familiar with the hydrides of the first two elements of Group VI, H_2O and H_2S. The hydride of the third member of this group, H_2Se, is usually not as well known. Based on your general chemical knowledge and periodic correlations, predict as much as you can about the chemistry and structure of H_2Se.

5. Carbon and silicon are both in the same group of the periodic table and both form oxides with similar empirical formulas, CO_2 and SiO_2. Despite this, these two compounds are quite different from each other. Describe some of the major differences between CO_2 and SiO_2, and explain why these differences exist.

PRACTICE PROBLEM SOLUTIONS

1. a. $H_2(g)$ + $F_2(g)$ → 2 HF(g)

                                 heat
   b. $H_2O(g)$ + $CO(g)$ → $H_2(g)$ + $CO_2(g)$
Depending on conditions, CH_3OH may be formed.

   c. 2 Na(s) + $H_2(g)$ → 2 NaH(s)

398

d. $Na(g)$ + $KCl(\ell)$ → $K(g)$ + $NaCl(\ell)$

e. $2\ Na(s)$ + $Cl_2(g)$ → $2\ NaCl(g)$

f. $2\ Na(s)$ + $O_2(g)$ → $Na_2O_2\ (g)$
$3000°C$

g. $SiO_2(s)$ + $3\ C(s)$ → $SiC(s)$ + $2\ CO(g)$

h. $2\ KClO_3(s)$ → $2\ KCl(s)$ + $3\ O_2(g)$

i. $Cl_2(g)$ + $2\ OH^-$ (cold aq)
→ $OCl^-(aq)$ + $Cl^-(aq)$ + $H_2O(\ell)$

2. Consult the table of reduction potentials found in Chapter 21. Those metal half reactions that are below the hydrogen half reaction would be good choices. Examples of such metals might be aluminum, zinc, calcium, sodium, and potassium. (Notice that the last two would be extremely reactive.)

3. a. $N_2(g)$ + $3\ H_2(g)$ → $2\ NH_3(g)$

b. $2\ NaCl$ + $CaCO_3$ → Na_2CO_3 + $CaCl_2$

c. $2\ NaCl(s)$ + $H_2SO_4(\ell)$ → $Na_2SO_4(s)$ + $2\ HCl(g)$

4. VSEPR Theory would suggest that H_2Se should have a bent structure with a bond angle somewhat smaller than the normal tetrahedral angle ($109.5°$). The electronegativity of selenium is much less than that of oxygen, and this suggests that hydrogen bonding will not be important in hydrogen selenide. This would lead one to suggest that hydrogen selenide is a gas under normal conditions of temperature and pressure. By analogy with hydrogen sulfide, you should expect hydrogen selenide to be a weak acid, and in fact it should probably be somewhat stronger than hydrogen sulfide.

5. (L2) Carbon dioxide is a simple, gaseous molecule with C=0 double bonds; silicon dioxide occurs is several different forms, but each form consists of infinite arrays of SiO_4 tetrahedra

sharing corners. Because of this network structure, silicates are solids under normal conditions.

PRACTICE TEST (30 Minutes)

1. (**L3**) Name each of the following.

a. $NaClO_2$ _____   b. $HOCl$ _____

c. N_2H_4 _____   d. $NaHCO_3$ _____

2. Complete and balance the following equations:

$$\text{a. } 2 H_2(g) + CO(g) \xrightarrow{\text{catalyst}}$$

$$\text{b. } 3 CaO(s) + 2 Al(s) \rightarrow$$

$$\text{c. } 2 NaCl(aq) + 2 H_2O(\ell) \xrightarrow{\text{electrolysis}}$$

$$\text{d. } 4 Al(s) + 3 O_2(g) \rightarrow$$

$$\text{e. } SiO_2(s) + 4 HF(\ell) \rightarrow$$

$$\text{f. } 3 Mg(s) + N_2(g) \rightarrow$$

$$\text{g. } NH_4NO_3(s) + heat(250°C) \rightarrow$$

$$\text{h. } Mg(s) + 2 HCl(g) \rightarrow$$

3. According to the text, hydrogen forms three different types of compounds. Name these three different types of compounds, and explain what type of elements are most likely to react with hydrogen and form these types of hydrides.

4. (**L1 and L2**) Write balanced chemical reactions for one method of preparation for each substance listed.
a. very pure silicon
b. nitric acid
c. oxygen gas
d. hydrazine

5. **(L1)** In 1966 a new compound was prepared consisting only of phosphorus and fluorine and having the empirical formula PF_2. It was observed that a sample of this compound having a mass of 0.217 grams would exert a pressure of 244 mmHg at $25^{\circ}C$ in a container having a volume of 120 mL. What is the molecular formula of this compound?

6. List at least five different examples of silicate minerals.

CONCEPT TEST ANSWERS

1. hydrogen   2. noble gases
3. ammonia   4. +1
5. kerosene or mineral oil   6. sea water
7. $NaHCO_3$ -baking soda   CaO - lime,

    Na_2CO_3 - soda ash or washing soda

8. Ca^{2+} and Mg^{2+} 9. +1,+2
10. calcium carbonate   11. bauxite
12. Because it's soft and weak unless alloyed with other elements.
13. aluminum oxide, Al_2O_3 14. silicon
15. silica or silicon dioxide
16. d.- NO, c.- N_2O_4, a.- N_2O, and b.- NO_2
17. oxygen   18. ozone, O_3
19. sulfuric acid   20. chlorine

PRACTICE TEST ANSWERS

1. **(L5, L6, and L8)**
a. sodium chlorate   b. hypochlorous acid
c. hydrazine   d. sodium bicarbonate

2. **(All sections)**   catalyst
    a. $2 H_2(g)$ + CO(g) $\rightarrow$ $CH_3OH(\ell)$

    b. 3 CaO(s) + 2 Al(s) $\rightarrow$ Al_2O_3(s) + 3 Ca(s)

    c.   2 NaCl(aq) + 2 $H_2O(\ell)$
           electrolysis
              $\rightarrow$ 2 NaOH(aq) + H_2(g) + Cl_2(g)

    d. 4 Al(s)   + 3 O_2(g) $\rightarrow$ 2 Al_2O_3(s)

e. $SiO_2(s) + 4 HF(\ell) \rightarrow SiF_4(g) + 2 H_2O(\ell)$

f. $3 Mg(s) + N_2(g) \rightarrow Mg_3N_2(s)$

g. $NH_4NO_3(s) + heat(250°C) \rightarrow N_2O(g) + 2 H_2O(g)$

h. $Mg(s) + 2 HCl(g) \rightarrow MgCl_2(s) + H_2(g)$

3. **(L2)** The three types are (a) anionic hydrides, formed when hydrogen reacts with metals of low electronegativity, (b) covalent hydrides, formed with electronegative elements, and (c) interstitial metallic hydrides, when the small hydrogen atoms are absorbed into the holes in the crystal lattice of a metal.

4. **(L5, L6, and L7)**
a. $SiO_2(s) + 2 C(s) \xrightarrow{heat} Si(\ell) + 2 CO(g)$

$Si(s) + 2 Cl_2(g) \rightarrow SiCl_4(\ell)$

$SiCl_4(g) + 2 Mg(s) \rightarrow 2 MgCl_2(s) + Si(s)$

b. $2 NaNO_3(s) + H_2SO_4(aq) \rightarrow 2 HNO_3(aq) + Na_2SO_4(s)$

An alternate method of synthesis, called the Ostwald Process, could also be used to answer this question.

b.
$NH_3(g) + 5/4 O_2(g) \xrightarrow{Pt} NO(g) + 3/2 H_2O(g)$

$NO(g) + 1/2 O_2(g) \rightarrow NO_2(g)$

$NO_2(g) + 1/3 H_2O(\ell) \rightarrow 2/3 HNO_3(aq) + 1/3 NO(g)$

c. Oxygen is normally prepared by fractional distillation of air, but it can also be chemically prepared by the reaction

$2 KClO_3(s) \xrightarrow{catalyst \& heat} 2 KCl(s) + 3 O_2(g)$

d. (The Raschig Process)
$2 NH_3(aq) + NaOCl(aq) \rightarrow N_2H_4(aq) + NaCl(aq) + H_2O(\ell)$

402

5. (**L6**) molar mass = 138 g/mole; formula = P_2F_4

6. (**L5**) Portland cement and olivine are examples of orthosilicates; the asbestos minerals are members of the group called amphiboles, and kaolinite is the example mentioned in the text to represent the large number of different clay minerals that are known. Feldspars and zeolites are examples of aluminosilicates.

CHAPTER 23
THE TRANSITION ELEMENTS

CHAPTER OVERVIEW

Looking at the transition metals is somewhat like looking at a history lesson. This group of elements includes some, like gold, silver, iron and copper, that were important very early in the development of human society and continue to play significant roles. Others, like platinum and its neighbors, have become important in more recent history but are now most useful, for example, as catalysts for industry. Despite the suggestions that this is the plastic age, metals are still essential for many activities.

One of the more interesting aspects of transition metal chemistry is the colorful compounds that these elements produce. Both natural materials, like gemstones, and artificial products, like glass and pigments, owe their colors to the electronic configurations of the transition metal ions. The study of these elements should certainly be popular with anyone who is interested in strong colors and unusual chemistry.

LEARNING GOALS

1. Understand the special chemical and physical properties of the transition metals, especially those that are related to the electronic configurations of these elements. (Sec. 23.1)

2. Be familiar with the commercial methods used for the production and purification of the transition metals. (Sec. 23.2)

3. Know how to name coordination compounds and also to identify the possible types of isomerism that may occur in this class of compounds. (Secs. 23.3 & 23.4)

4. Be able to use a modern theory of chemical bonding to discuss the bonding in coordination compounds and to explain the magnetic properties

404

and colors of coordination compounds. (Secs. 23.4 & 23.6)

ARMCHAIR EXERCISE

1. On the periodic table, the elements zinc and scandium appear to be in the transition metal group, but most compounds of these elements are not colored. Can you suggest an explanation for this behavior?

CONCEPT TEST

1. _____ and _____ are the most commonly observed oxidation numbers in compounds of the first transition series.

2. Higher oxidation states are _____ (choose more or less) common for compounds of the elements in the second and third transition series.

3. Several observations, including the fact that radii of the sixth period transition elements are almost identical to those of the corresponding fifth period transition elements, are explained by the _____.

4. Most of the d-block and f-block elements have unpaired electrons in the ground state, and their magnetic properties are said to be _____.

5. _____ is the use of high temperatures to recover metals from their ores, and

_____ is the use of low temperature, aqueous solutions for the same purpose.

6. The combination of calcium silicate and metal oxides that floats on top of the liquid iron in a blast furnace is called _____.

7. In some cases, metal ores are roasted, that is, they are heated in the presence of _____ gas so that the metal is converted to the metal oxide.

8. The compounds called _____ consist of a metal ion associated with a group of neutral molecules or anions, such as $Cu(NH_3)_6^{2+}$.

9. Molecules or ions bonded to the central metal ion in a coordination compound are called _____.

10. When a ligand has more than one Lewis base site, it can bond to a metal ion in each of these sites to form a type of coordination compound called a(n) _____.

11. In naming a coordinating compound, if the complex ion is an anion, the suffix _____ replaces the normal ending of the metal name.

12. Give the oxidation number of the metal ion in each complex.

a. $[PtCl_4]^{2-}$ _____ b. $[Co(NH_3)_4Cl_2]^+$ _____

13. Beside each prefix, write the corresponding numerical value.

a. tris _____ b. tetrakis _____ c. bis _____

14. The splitting of the _____ is the cause of both the magnetic behavior and the color for coordination compounds.

15. In naming a coordination compound, if the ligand is an anion with a name that normally ends in -ide, the ending is changed to _____.

16. _____ are molecules with the same stoichiometry but different atomic arrangements. When two isomers are nonsuperimposable mirror images of each other, the compounds are called

_____.

17. The _____ series results when ligands are arranged in order of their ability to cause orbital splittings.

STUDY HINTS

1. It was noted earlier that students sometimes confuse ammonia and ammonium. Also be careful to distinguish between the names ammine, which represents ammonia acting as a ligand, and amine, which is an organic functional group.

2. Remember that in coordination compounds, some of the species may be covalently bonded to the central metal ion, while others are bonded ionically. When such a compound is dissolved in water, the ligands

that are coordinated usually remain bonded to the central metal ion, and so are less likely to react. The brackets indicate which species are coordinated and which are not. This is important when trying to predict the reactions of coordination compounds.

3. Molecular models can be very helpful in the study of isomerism.

4. When naming coordination compounds, if the coordinated species is an anion the metal name must end in -ate. For a few metals the name is also changed to the Latin form. Some examples of this change are iron becoming ferrate, silver becoming argentate, and gold becoming aurate. Watch for examples of these changes.

5. If the name of a coordination compound contains a double vowel it is difficult to pronounce, and there is a temptation to drop one of the vowels. For example, hexaamminecobalt(III) ion could become hexamminecobalt(III). Watch for this when naming coordination compounds.

PRACTICE PROBLEMS

1. **(L1)** Give spectroscopic notation for the electronic configurations of the following transition metal ions and in each case indicate whether or not it is paramagnetic.

a. Zn^{2+} _____

b. Re^{2+} _____

c. Mn^{2+} _____

d. Zr^{4+} _____

2. **(L3)** Name the following ligands:

a. Br^- _____   b. CN^- _____

c. en _____   d. NCS^- _____

e. F^- _____   f. CO _____

3. **(L3)** Name the following coordination compounds

a. $[Ti(H_2O)_6]^{3+}$ _____

b. $[AgF_4]^-$ _____

c. $[Co(CO)_4]^-$ _____

d. $[Pt(NH_3)_2Br_2]$ _____

e. $[Cu(NH_3)_4]SO_4$ _____

f. $[W(CN)_8]^{3-}$ _____

4. **(L3)** Write the formulas for each of the following compounds or ions:

a. tris(ethylenediamine)cobalt(III) _____

b. tris(oxalato)ferrate(III) _____

c. hexaamminecobalt(III) ion _____

d. hexafluoroantimonate(V) ion _____

5. **(L3)** Draw the geometric isomers of the following compounds and label each as a cis or trans isomer.

a. $[Co(NH_3)_4Br_2]^+$ b. $[Pt(NH_3)_2Br_2]$

6. **(L2)** Explain the fundamental chemistry of a basic oxygen furnace, using a few typical reactions to demonstrate the discussion.

PRACTICE PROBLEM SOLUTIONS

1. a. $[Ar]3d^{10}$ diamagnetic
   b. $[Xe]4f^{14}5d^{5}$ paramagnetic
   c. $[Ar]3d^{5}$ paramagnetic
   d. $[Kr]4d^{0}$ diamagnetic

2. a. Br^{-} bromo
   b. CN^{-} cyano
   c. en   ethylenediamine
   d. NCS^{-} thiocyanato
   e. F^{-} fluoro
   f. CO   carbonyl

3. a. hexaaquqtitanium(III)
   b. tetrafluoroargentate(III)
   (See helpful hint #4.)
   c. tetracarbonylcobalt(-1)
   d. diamminedibromoplatinum(II)
   e. tetraamminecopper(II) sulfate
   f. octacyanotungsten(V)

4. a. $[Co(en)_3]^{3+}$
   b. $[Fe(C_2O_4)_3]^{3-}$
   c. $[Co(NH_3)_6]^{3+}$
   d. $[SbF_6]-$

5. a.

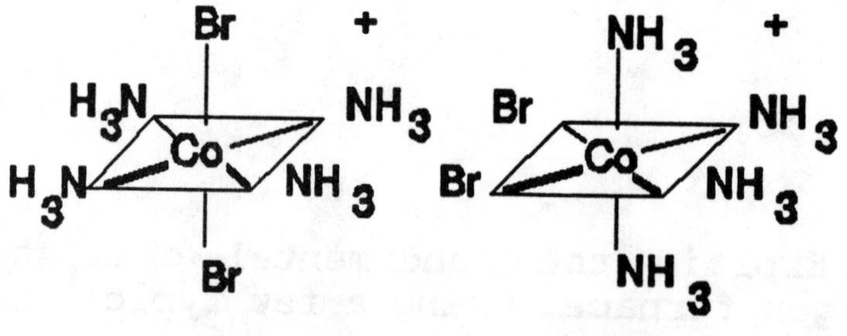

trans isomer   cis isomer

b.

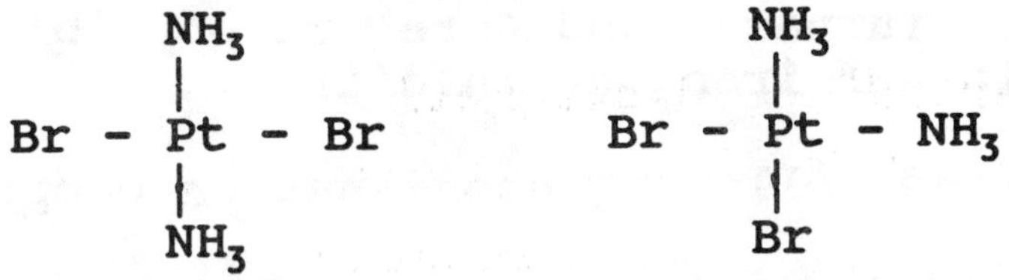

$$Br - \underset{\underset{NH_3}{|}}{\overset{\overset{NH_3}{|}}{Pt}} - Br \qquad\qquad Br - \underset{\underset{Br}{|}}{\overset{\overset{NH_3}{|}}{Pt}} - NH_3$$

trans isomer   cis isomer

6. The major purpose of a basic oxygen furnace is to remove nonmetallic impurities from pig iron. When oxygen gas is blown through the molten iron, phosphorus, sulfur, and carbon are oxidized, as shown in the following reactions:

$$4\ P(s) \qquad +\ 5\ O_2(g) \qquad \rightarrow\quad 2\ P_2O_5(s)$$

$$C(s) \qquad +\ O_2(g) \qquad \rightarrow\quad CO_2(g)$$

$$S(s) \qquad +\ O_2(g) \qquad \rightarrow\quad SO_2(g)$$

These acidic oxides react with basic oxides, such as CaO, that are added or used to line the furnace.

$$P_2O_5(s) \qquad +\ 3\ CaO(s) \qquad \rightarrow\quad Ca_3(PO_4)_2(s)$$

The silicate minerals present, called gangue, are removed by reacting with the lime that results from heating the limestone.

$$SiO_2(s) \qquad +\ CaO(s) \qquad\qquad \rightarrow\quad CaSiO_3(s)$$

PRACTICE TEST (30 Min.)

1. Write the electronic configuration for each of the following transition metal ions and indicate in each case whether or not the ion is paramagnetic.

a. Mo^{3+} b. Co^{2+} c. Cd^{2+} d. Cu^+

2. For each of the following compounds draw all of the possible geometric isomers and stereoisomers.

a. $[Co(en)Br_4]^-$ b. $[Cr(NH_3)_4Br_2]^+$ c. $[Co(en)_3]^{3+}$

3. Discuss the sequence of reactions that occur in a blast furnace and are responsible for the conversion of iron ore into iron.

4. Name the following coordination compounds

a. $[Ag(NH_3)_2]^+$

b. $[Co(en)_3]_2(SO_4)_3$

c. $K_2[MnF_6]$

d. $[Co(NH_3)_5Cl]Cl_2$

e. $[SbCl_6]^{3-}$

f. $K_2[NiF_6]$

5. Write the formulas for each of the following compounds or ions:

a. tetrabromoplatinum(II) ion
b. hexaaquairon(III) ion
c. potassium diiodoargentate(I)
d. tetraaquadichlorochromium(III) ion

6. A certain coordination compound has the empirical formula $Cr(NH_3)_4Cl_3$, but when silver nitrate is added to a solution of this species, only one mole of silver chloride is formed per mole of chromium in the solution. If possible, suggest an explanation for this observation.

7. Gold and platinum are among the most dense elements known. Suggest an explanation in terms of atomic structure for this observation.

COMMENTS ON ARMCHAIR EXERCISE

1. For coordination compounds, the color usually results from the splitting of the set of d orbitals. Electrons in these orbitals shift from one set to another, and the energy changes correspond to the colors we observe. The common oxidation states of these elements, Scandium(III) and Zinc(II), have no d electrons and so would not be expected to produce color in this way.

CONCEPT TEST ANSWERS

1. +2 and +3

2. more

3. lanthanide contraction   4. paramagnetic
5. pyrometallurgy, hydrometallurgy
6. slag   7. oxygen
8. coordination compounds   9. ligands
10. chelate   11. -ate
12. platinum(II), cobalt(III)
13. a. tris 3   b. tetrakis 4   c. bis 2
14. d-orbitals   15. -o
16. isomers, optical isomers
17. spectrochemical

PRACTICE TEST ANSWERS

1. **(L1)** a. $[Kr]4d^3$ paramagnetic

   b. $[Ar]3d^7$ paramagnetic

   c. $[Kr]4d^{10}$ diamagnetic

   d. $[Ar]3d^{10}$ diamagnetic

2. **(L3)** a. Neither cis-trans nor stereoisomers are possible for this compound, that is, it exists in only one form.

b. This compound does have a pair of cis-trans isomers, as shown below.

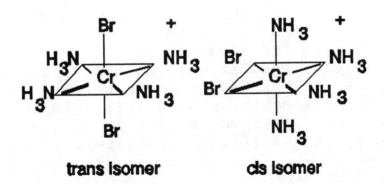

trans isomer   cis isomer

c. This compound has no cis-trans isomers, but it does have a pair of stereoisomers, as shown below.

413

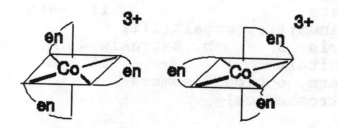

3. **(L2)** The coke burns and produces carbon monoxide.

$$2 \, C(s, \text{ coke}) + O_2(g) \rightarrow CO(g)$$

The carbon monoxide reduces the iron oxide to the metal.

$$Fe_2O_3(s) + 3 \, C(s) \rightarrow 2 \, Fe(\ell) + 3 \, CO_2(g)$$

Much of the carbon dioxide produced is reduced by the coke to form more carbon monoxide for the above reaction.

$$CO_2(g) + C(s) \rightarrow 2 \, CO(g)$$

4. **(L3)** a. diamminesilver(I) ion
   b. tris(ethylenediamine)cobalt(III) sulfate
   c. potassium hexafluoromanganate(IV)
   d. pentaamminechlorocobalt(III) chloride
   e. hexachloroantimonate(III) ion
   f. potassium hexafluoronickelate(IV)

5. **(L3)** a. $[PtBr_4]^{2-}$

   b. $[Fe(H_2O)_6]^{3+}$

   c. $K[AgI_2]$

   d. $[Cr(H_2O)_4Cl_2]^+$

414

6. **(L3)** The chloride ions that are coordinated to the cobalt will not precipitate, but those outside the coordination sphere will. This suggests that the reaction is

$$[Co(NH_3)_4Cl_2]Cl(aq) + Ag^+(aq) \rightarrow [Co(NH_3)_4Cl_2]^+ + AgCl(s)$$

7. **(L4)** The lanthanide contraction causes the radii of transition elements in the sixth period to be about the same size as the their counterparts in the fifth period. This relatively small size and the higher mass causes the densities to be unusually high.

CHAPTER 24
NUCLEAR CHEMISTRY

CHAPTER OVERVIEW

There are many practical applications of nuclear chemistry, and in addition, exposure to natural sources of radiation is an inescapable fact of life. Indeed, radiation is encountered so commonly, that it's difficult to believe that it was only about a century ago that scientists recognized and began to understand radioactivity.

As more has became known about radioactivity, it has become possible to use this information to create specific radioisotopes and design specific reactions. Sometimes the results of these efforts have been extremely beneficial, such as medical diagnostics and radiochemical dating, but in other cases, the results have been more controversial.

Both nuclear fission and nuclear fusion were developed based on the knowledge of the behavior of radioactive elements. Continued development in this field will probably reveal new opportunities for both good or bad uses of the resulting discoveries.

LEARNING GOALS

1. Be able to write nuclear reactions for the various types of spontaneous radioactive processes described. (Sec. 24.2)

2. Be familiar with the factors that are important in determining nuclear stability and understand the importance of binding energy in the determination of nuclear stability. (Sec. 24.3)

3. Be able to do calculations involving the relationship between elapsed time and the amount of radioactive material which has been transformed, using the equation

$$\ln \frac{N}{N_o} = -kt$$

This equation is also the basis for radiochemical dating calculations. (Sec. 24.4)

4. In addition to natural radioactive processes, you should also be able to write equations for artificial nuclear transmutations, including nuclear fission and nuclear fusion. (Secs. 24.5, 24.6, and 24.7)

5. Be familiar with some commercial applications of radiation in medicine, chemical research, and chemical analysis. (Sec. 24.8 and 24.9)

ARMCHAIR EXERCISES

1. The enriched uranium pellets used in some nuclear reactors can be held in the bare hand with relatively little danger before they are placed in the reactor, but after the reactor has been running for a short time, those same pellets are so dangerous that even a short exposure would be harmful. Can you explain this difference?

2. The hydrogen fusion bomb uses a fission-type bomb as a trigger to initiate the reaction. What does this tell you about the activation energy of the fusion reaction? As the text notes, the material that undergoes fusion in a hydrogen bomb is actually lithium deuteride, LiD. Can you think of any reasons why this substance is used instead of some other deuterated compound, such as deuterated hydrogen, D_2, or deuterated water, D_2O?

CONCEPT TEST

1. Beta radiation consists of _____.

2. Alpha radiation consists of _____.

3. In order to have a correct nuclear reaction the total number of _____ must remain the same.

4. _____ radiation has the greatest ability to penetrate matter.

5. In a correct equation for a nuclear reaction, the sum of the mass numbers of the reacting nuclei must equal the sum of the _____

6. In a correct nuclear equation, the sum of the atomic numbers of the products must equal the sum of _____ .

7. Another type of radioactive decay consists of the emission of a positively charged electron, a particle called a(n) _____ .

8. All isotopes having an atomic number greater than that of the element _____ are radioactive.

9. List four stable isotopes that have both an odd number of protons as well as an odd number of neutrons. _____

10. The point of maximum stability in the binding energy curve occurs in the vicinity of what isotope of what element? _____

11. What effect does temperature have on the rate of radioactive decay? _____

12. Radiochemical dating is based on the assumption that the _____ of each radioactive isotope was the same in ancient times as it is now.

13. The _____ of a radioactive sample is a measure of the rate of nuclear decay and can be measured using a Geiger counter.

14. If a nuclear fission reaction produces more neutrons than would be necessary to cause another fission reaction to occur, the result can be a sequence of rapidly occurring reactions called a(n) _____ .

15. When light nuclei combine to form heavier nuclei, releasing large amounts of energy, the process is called _____.

16. Approximately _____ % of the electric power in the United States is produced by nuclear reactors. (Circle the best answer.)

a. 10%   b. 20%   c. 30%   d. 40%   e. 50%

17. Nuclear fusion reactions of the element _____ are the primary source of energy from the sun.

18. The unit called the _____ is a measure of radiation exposure and is proportional to the amount of ionization produced in air by x-rays and gamma rays.

19. The unit called the _____ measures the radiation dose to tissue rather than air.

20. Radioactive isotopes are used in nuclear medicine in two different ways, _____ and _____.

STUDY HINTS

1. Be sure that you can identify the common types of nuclear radiation, that is, alpha, beta, and gamma (α, β, and γ) by name, by appropriate symbol, and in terms of the nature of the radiation. Questions on nuclear reactions may list the type of radiation emitted either by symbol or by description.

2. Your instructor may wish you to know some of the terminology used to describe radioactive exposure, such as Curie, Roentgen, rad, and rem. Listen carefully to determine if this is required.

3. Remember that there are two different types of logarithms on your calculator, base 10 logs (usually designated **log**) and natural logs

(designated **ln x**). You will use the latter type of logs for these problems; try to avoid confusing the two sets of calculator keys.

4. You must read radioactive decay problems carefully. The quantity, N, in the equation is the amount of material <u>remaining</u>. Sometimes a problem will refer to the amount of matter that has reacted. Be aware of this possible source of confusion as you are working problems, and watch for this situation on examinations.

PRACTICE PROBLEMS

1. **(L1 and L4)** Fill in the single missing species to complete each reaction.

a. $^{234}_{90}\text{Th}$ $\rightarrow$ ____ $+$ $^{230}_{88}\text{Ra}$

b. $^{11}_{5}\text{B}$ $+$ $^{4}_{2}\text{He}$ $\rightarrow$ $^{1}_{0}\text{n}$ $+$ _____

c. ____ $\rightarrow$ $^{0}_{-1}\beta$ $+$ $^{127}_{53}\text{I}$

d. $^{131}_{56}\text{Ba}$ $+$ $^{0}_{-1}\beta$ $\rightarrow$ ____

e. $^{27}_{13}\text{Al}$ $+$ $^{3}_{1}\text{H}$ $\rightarrow$ ____ $+$ $^{27}_{12}\text{Mg}$

f. $^{2}_{1}\text{H}$ $+$ ____ $\rightarrow$ $^{4}_{2}\text{He}$ $+$ $^{1}_{0}\text{n}$

g. $^{40}_{19}\text{K}$ $\rightarrow$ $^{40}_{18}\text{Ar}$ $+$ ____

h. $^{55}_{26}\text{Fe}$ $+$ ____ $\rightarrow$ $^{55}_{25}\text{Mn}$

2. **(L2)** Each of these pairs of nuclides consists of one stable isotope and one that is unstable. Select the isotope that is more likely to be stable in each case and explain your choice.

a. $^{96}_{43}\text{Tc}$ or $^{96}_{42}\text{Mo}$ b. $^{14}_{7}\text{N}$ or $^{30}_{15}\text{P}$

3. (L3) Underwater archaeologists discovered a sunken ship off the coast of Israel just south of Haifa. The ship's cargo included an amazingly well-preserved cargo of lead and copper ingots and is thought to have sunk during the Late Bronze Age (1550-1200 B.C.). Radiocarbon dating on wood from this ship shows a carbon-14/carbon-12 ratio that is 0.673 times that for comparable wood from the present time. If the half-life of carbon-14 is 5730 years, does the observed radiocarbon date agree with the suspected age of the ship?

4. (L1) a. One of the well-known radioactive decay series begins with uranium-238 and proceeds by emission of the following sequence of particles: alpha, beta, beta, alpha, alpha, alpha, and alpha. What is the isotope formed at conclusion of this set of processes?

b. Another well-known radioactive decay series begins with thorium-232 and proceeds by emission of the following sequence of particles: alpha, beta, beta, alpha, alpha, alpha, and alpha. What is the isotope formed at the conclusion of this set of processes?

5. (L2) Calculate the binding energy in kilojoules per mole of boron for the formation of boron-10. You will need to know the following masses: $_1^1H = 1.00783$; $_0^1n = 1.00867$; and $_5^{10}B = 10.01294$

PRACTICE PROBLEM SOLUTIONS

1. a. mass number: $234 = A + 230$, thus $A = 4$
      atomic number: $90 = Z + 88$, thus $Z = 2$
      the missing species is helium.

   $$_{90}^{234}Th \rightarrow \ _2^4He \ + \ _{88}^{230}Ra$$

   b. mass number: $11 + 4 = 1 + A$, thus $A = 14$
      atomic number: $5 + 2 = 0 + Z$, thus $Z = 7$
      the missing species is nitrogen.

   $$_5^{11}B \ + \ _2^4He \ \rightarrow \ _0^1n \ + \ _7^{14}N$$

   c. mass number: $A = 0 + 127$, thus $A = 127$
      atomic number: $Z = -1 + 53$, thus $Z = 52$
      the missing species is tellurium.

   $$_{52}^{127}Te \ \rightarrow \ _{-1}^0\beta \ + \ _{53}^{127}I$$

   d. mass number: $131 + 0 = A$, thus $A = 131$
      atomic number: $56 - 1 = Z$, thus $Z = 55$
      the missing species is cesium.

   $$_{56}^{131}Ba \ + \ _{-1}^0\beta \ \rightarrow \ _{55}^{131}Cs$$

   e. mass number: $27 + 3 = A + 27$, thus $A = 3$
      atomic number: $13 + 1 = Z + 12$, thus $Z = 2$
      the missing species is helium.

   $$_{13}^{27}Al \ + \ _1^3H \ \rightarrow \ _2^3He \ + \ _{12}^{27}Mg$$

f. mass number: $2 + A = 4 + 1$, thus $A = 3$
   atomic number: $1 + Z = 2 + 0$, thus $Z = 1$
   the missing species is hydrogen.

$$^2_1H + ^3_1H \rightarrow ^4_2He + ^1_0n$$

g. mass number: $40 = 40 + A$, thus $A = 0$
   atomic number: $19 = 18 + Z$, thus $Z = 1$
   the missing species is a positron.

$$^{40}_{19}K \rightarrow ^{40}_{18}Ar + ^0_1\beta$$

h. mass number: $55 + A = 55$, thus $A = 0$
   atomic number: $26 + Z = 25$, thus $Z = -1$
   the missing species is an orbital electron.

$$^{55}_{26}Fe + ^0_{-1}\beta \rightarrow ^{55}_{25}Mn$$

2. a. Molybdenum-96 has an even number of protons and neutrons. Technetium-96 has odd numbers of both protons and neutrons.

b. Nitrogen-14 is one of the few isotopes that is stable despite having odd numbers of both protons and neutrons. Phosphorus-30 is not one of those exceptions.

3. First it is necessary to determine the value of the rate constant for the radioactive decay of carbon-14.

$$k = \frac{0.693}{t_{1/2}} = \frac{0.693}{5730 \text{ yr}} = 1.209 \times 10^{-4} \text{ yr}^{-1}$$

Now substitute in the first order rate equation.

$$\ln \frac{N}{N_o} = -kt$$

$$\ln (0.673) = -(1.209 \times 10^{-4} \text{ yr}^{-1}) \times t$$

$$\underline{t = 3270 \text{ yr}}$$

This places the age of the ship within the range of dates given for the Late Bronze Age.

4. **a.** Each alpha particle represents the loss of four units of mass; each beta particle represents the loss of zero units of mass. Since five alpha particles were emitted, the total mass change is

$$\text{mass change} = 5 \times -4 = -20 \text{ amu}$$

Each alpha particle represents a decrease of two in the atomic number; each beta particle represents a gain of one unit in atomic number. Since five alpha particles and two electrons were emitted the total change in atomic number is

atomic number change = 5 x (-2) + 2 x 1 = -8

The new mass number is 238 - 20 = 218

The new atomic number is 92 - 8 = 84

<u>Thus the product at this stage is $^{218}_{84}\text{Po}$</u>

b. Since five alpha particles were emitted, the total mass change is

$$\text{mass change} = 5 \times -4 = -20 \text{ amu}$$

Since five alpha particles and two electrons were emitted the total change in atomic number is

atomic number change = 5 x (-2) + 2 x 1 = -8

The new mass number is 232 - 20 = 212

The new atomic number is 90 - 8 = 82

<u>The product at this stage is $^{212}_{82}\text{Pb}$</u>

5. Boron-10 consists of 5 protons and 5 neutrons, so the total predicted mass should be

$$5 \times 1.00783 + 5 \times 1.00867 = 10.08250$$

The mass defect is the difference between the actual and the predicted mass.

Mass defect = actual mass - predicted mass

Mass defect = 10.01294 g/mol - 10.08250 g/mol

Mass defect = - 0.06956 g/mol (or 6.956×10^{-5} kg)

Energy liberated = mass defect x c^2

$$= - 6.956 \times 10^{-5} \text{ kg/mol} \times (2.998 \times 10^8 \text{ m/s})^2$$

<u>Energy</u> $= - 6.26 \times 10^{12}$ J/mol

PRACTICE TEST (30 Min.)

1. Fill in the single missing species to complete each reaction.

a. $^{238}_{92}\text{U}$ $\rightarrow$ $^{4}_{2}\text{He}$ + _____

b. $^{234}_{90}\text{Th}$ $\rightarrow$ _____ + $^{234}_{91}\text{Pa}$

c. $^{96}_{42}\text{Mo}$ + $^{2}_{1}\text{H}$ $\rightarrow$ _____ + $^{1}_{0}\text{n}$

d. $^{1}_{0}\text{n}$ + $^{235}_{92}\text{U}$ $\rightarrow$ $^{143}_{55}\text{Cs}$ + $2\,^{1}_{0}\text{n}$ + _____

e. $^{59}_{26}\text{Fe}$ $\rightarrow$ _____ + $^{59}_{27}\text{Co}$

f. $^{12}_{6}\text{C}$ + $^{12}_{6}\text{C}$ $\rightarrow$ $^{1}_{1}\text{H}$ + _____

g. $^{40}_{19}\text{K}$ $\rightarrow$ _____ + $^{0}_{+1}\beta$

2. Each of these pairs of nuclides consists of one stable isotope and one that is unstable. Select the isotope that is more likely to be stable in each case and explain your choice.

a. $^{11}_{6}\text{C}$ or $^{12}_{6}\text{C}$ b. $^{238}_{92}\text{U}$ or $^{197}_{79}\text{Au}$

3. Strontium-90, a product of the explosion of nuclear fission bombs, has a half-life of 28.8 years. (a) What fraction of the strontium-90 produced by an atmospheric nuclear test will still

remain somewhere in the environment after a period of 50.0 years? (b) How long must pass before 90.0% of an initial amount of strontium-90 has vanished?

4. Match the radioisotope on the left with the appropriate letter for its use from the right hand column.

_____ cobalt-60   a. used to treat and diagnose thyroid gland disorders

_____ iodine-125   b. a fissionable nuclei

_____ carbon-14   c. used to irradiate food

_____ plutonium-239 d. used to "tag" compounds

5. If a 10.0 grams sample of the radioactive isotope cesium-134 is allowed to remain undisturbed for 1.6 years, only 6.00 grams of the cesium is found to remain. What is the half-life of cesium-134?

COMMENTS ON ARMCHAIR EXERCISES

1. Naturally occurring uranium isotopes have very long half lives. What does this tell you about how rapidly these atoms undergo radioactive decay? In turn, what does this tell you about the amount of radiation that will be emitted during a short time by a small sample? On the other hand, once the uranium has been bombarded by neutrons in a nuclear reactor, some of it has been transformed into elements having much shorter half lives. Why are these more dangerous?

2. The activation energy for the fusion reaction is, as you have surely guessed, very high. One problem this causes is that the high temperature necessary for fusion causes the atoms to expand so rapidly that the possibilities for atomic collisions rapidly decrease. The longer that the material to undergo fusion remains in a condensed state, the more likely the fusion reaction will become, even if the amount of time involved is very small. Which of the three possible compounds has

the highest boiling point and so will remain in the more dense form for the longest time when the temperature begins to rise rapidly?

The lithium salt also has one more advantage; it can contribute to the fusion explosion. The lithium may undergo a fusion reaction (see the binding energy curve on page 1096) or it may absorb a neutron, then break apart, forming helium and hydrogen atoms that can undergo fusion.

CONCEPT TEST ANSWERS

1. high energy electrons   2. helium nuclei
3. nucleons (protons plus neutrons) 4. gamma (γ)
5. mass numbers of the nuclei produced
6. atomic numbers of the reactants.
7. positron   8. bismuth
9. $^{2}_{1}H$, $^{6}_{3}Li$, $^{10}_{5}B$, and $^{14}_{7}N$

10. iron-56   11. none
12. activity (or half-life)   13. activity
14. chain reaction   15. fusion
16. b. 20%   17. hydrogen
18. roentgen(R)   19. rad
20. diagnosis and therapy

PRACTICE TEST ANSWERS

1. (L1 and L4)

a. $^{238}_{92}U \rightarrow ^{4}_{2}He + ^{234}_{90}Th$

b. $^{234}_{90}Th \rightarrow ^{0}_{-1}\beta + ^{234}_{91}Pa$

c. $^{96}_{42}Mo + ^{2}_{1}H \rightarrow ^{97}_{43}Tc + ^{1}_{0}n$

d. $^{1}_{0}n + ^{235}_{92}U \rightarrow ^{143}_{55}Cs + 2\,^{1}_{0}n + ^{91}_{37}Rb$

e. $^{59}_{26}Fe \rightarrow ^{0}_{-1}\beta + ^{59}_{27}Co$

f. $^{12}_{6}C + ^{12}_{6}C \rightarrow ^{1}_{1}H + ^{23}_{11}Na$

g. $^{40}_{19}K \rightarrow {}^{40}_{18}Ar + {}^{0}_{+1}\beta$

2. (**L2**) a. Carbon-12 has an even number of both protons and neutrons, and so is favored over carbon-11, which has an even number of protons but an odd number of neutrons.

b. Gold-197 is more likely to be stable, since uranium-238 has an atomic number greater than that of bismuth (which is the element with the highest atomic number that does not undergo radioactive decay).

3. (**L3**) (a) The value of k is 2.41×10^{-2} yr^{-1}, and the fraction gone after 50.0 years would then be 0.300. (b) 95.7 years must pass before 90.0% of the strontium has decayed.

4. (**L5**) c. used to irradiate food   - cobalt-60
    a. used to treat and diagnose
       thyroid gland disorders   - iodine-125

    d. used to "tag" compounds   - carbon-14

    b. a fissionable nucleus   - plutonium-239

5. (**L3**) The rate constant is 0.32 yr^{-1} and the half-life is 2.2 years.